SOFTWARE PROJECT MANAGEMENT

*CREATING AND MANAGING
A SUCCESSFUL PLAN*

GORD TALLAS

outskirtspress
DENVER, COLORADO

To my parents who always were positive and provided support –
John and Hertha Tallas, and to my family,
my wife Marie, and my children Blake, Kiera and Riley.

TABLE OF CONTENTS

PREFACE

This book is based on real life Project Management of Information Technology and other business projects. It is based on experience in creating or approving well over 150 Project Plans. My formative years in Project Management were with a systems integration firm which developed custom applications, which were contractually constrained to a fixed price and fixed delivery date. This forced me to hone my detailed planning skills. Over the course of my career there, none of the projects I was associated with were over-budget and always were delivered on time.

It is my belief, that every organization can do this, if they have the proper focus. They must want the consistency and understand the dynamics of scope, duration and cost. Too often, internal and vendor plans are sparse (excel spreadsheets, tasks with 20+ days duration, etc), and very difficult to deliver successfully. Many organizations simply adjust their "plans" regularly as the project slips – and in the end, no one knows how much the delay costs the organization. If there is no detailed plan, it is unlikely the project will succeed. Senior executives can't stand these shifting plans, scope and budgets.

Too often, the cost of missing a deadline is not understood. Organizations that have proper plans, can assess the cost of missing a deadline by 1 month or 2 months in real dollars. It is only when there is knowledge about the cost of missing delivery dates will there be an impetus to utilize proper Project Management principles, and deliver high quality, realistic plans.

This book strives to deliver specific details on what needs to be done to create and manage a successful plan.

Good luck.

Gord Tallas

ACKNOWLEDGMENTS

I would like to take this opportunity to thank the many organizations and individuals I had the pleasure to work with during my career. In particular I would like to thank the executive management at the Great West Life (GWL) Assurance Company for providing me the opportunity to lay down and expand many of the ideas, thoughts and methodology. Nick Curry, the then CIO, was instrumental in providing not only this opportunity, but to provide constant support throughout the business units for this endeavor. I would also like to thank Ross Peterson and Patrick Smith for their help in reviewing the technical chapters and providing excellent feedback. Finally, I would like to thank all of the organizations and colleagues who have reviewed this book and provided their thoughtful commentary.

EXECUTIVE SUMMARY

1.1 WHAT IS PROJECT MANAGEMENT

Almost every day each of us does some facet of Project Management. Take Saturday morning when we list the chores we need to do, decide if we can accomplish all of them and if not scale down the list, decide on the order, and then proceed to do the first task. We've just done many of the fundamentals of Project Management:

- Define the Scope (task list)

- Estimate the Tasks

- Revise Scope

- Schedule Plan

- Execute the Plan

Many facets of Project Management of an Information System project are similar to managing any project, from managing the day's chores to building a house.

While there are many chapters that only relate to software engineering, the majority of this book is concerned with proper Project Management. Wherever there is a need to plan and manage an endeavor, these principles apply. For instance, these principles and methodology for planning and managing projects were successfully employed in a multi-million dollar project for conducting a due diligence for a possible strategic alliance. Twenty three plans were developed that addressed such facets as assessment of assets and liabilities, physical moving of people, equipment and files, transferring business units and people, training and communications. Of the 23 plans, only 4 were plans for conducting some facet of software conversion, analysis, design and construction. The remaining were plans that were concerned with assessing the business, and planning for the alliance.

Project Management has clearly been around for thousands of years. Yet the basics are often not learned or once learned not used. If you queried IS Project Managers today, you would find that far too many believe that Project Management is the weekly obtaining of project status from all team members, determining how things are going, creating management status reports, and adjusting the plan if there is an overrun. Often they do not even understand the management reports that they are creating. I call this project tracking, a component of project management but certainly not the most important component.

Webster's Dictionary defines "project" as:

1)   A specific plan or design

2)   Scheme; idea.

3)   A planned undertaking.

4)   A definitively formulated piece of research.

Webster's Dictionary defines "management" as:

1)   The act or art of managing.

2)   The conducting or supervising of something.

3)   Judicious use or means to accomplish an end.

4)   Capacity for managing, executive skill.

5)   The collective body of those who manage or direct an enterprise.

According to these definitions, "Project Management" could be "Judicious use or means to accomplish a planned undertaking". I like that, it is fairly good, especially the word "judicious". This implies that there are options, and the Project Manager is assessing the options and determining the "right" course of action to take.

Another definition may be summarized as "being able to craft a good plan to complete a project and then to execute the plan on time and on budget". You will notice that I have included the task of **planning** the plan and then executing it.

Given that definition, what is the main objective of the Project Manager? It should be:

> "To successfully deliver a project on time, on budget and to specification"

The underlying theme that is stated here, is that simply being "on time and on budget" is not enough. We must deliver what we set out to deliver, and in order for the project to be deemed successful, it must be of high quality. Delivering a system on time that is bug ridden is not acceptable.

There are five keys to accomplish this objective:

- Ensure there is a clear, well understood and complete definition of the scope and objectives of the system,

- Develop an overall project plan that has realistic estimates and is crafted using good project management principles,

- Control **all** scope changes through the use of rigorous change control procedures,

- Ensure all team members have the correct skill set and have no obstructions to completing their tasks as planned, and

- Ensure the end User is 100% committed to the initial plan and the on-going execution of the plan.

The objective and keys to achieving the objective sound simple enough don't they? Aren't all projects like this? Alas, no. This document is intended to provide detailed information on how to accomplish the objective noted above, via the rigorous project management principles that address these key points.

1.2   DOCUMENT AUDIENCE

This document is intended for a wide range of audiences, including systems personnel, the user community, and non-information systems Project Managers.

The primary intent is to provide Project Managers with the methodology required to plan and successfully deliver projects. However, analysts, programmers, and other system professionals all participate to varying degrees in estimating, planning, organizing and controlling projects. The ideas herein should be of interest to all systems professionals. For non-IS Project Managers, the following chapters can be scanned rather than read since they primarily relate to information systems concepts: Chapters 5, 6, 7, 17, and 18 and Appendices A, B and C.

In addition, some members of the User community may find this document useful. In particular, the "User Coordinator" who is responsible for the planning and coordinating of all User activities and User personnel on the project will be able to use a range of the ideas described herein. A casual User interested in Project Management should read Chapters 1 through 4, 8 through 15 and Appendix C.

1.3   DOCUMENT ORGANIZATION

Chapter 1 of this document is the overview. It presents an overview of the purpose and structure of this document.

Chapter 2 of this document presents **overall** project management principles

Chapters 3 through 9 of this document present the principles of the **Planning** component of project management.

Chapters 10 through 14 present the principles for **Delivering** a project.

Chapter 15 presents the principles of the **Review** component of project management.

Chapter 16 documents the roles and responsibilities of the Project Manager.

Chapter 17 defines four system components, Business Process Re-Engineering (BPR), Enterprise Wide Architecture (EWA), Technology Selection and Deployment (TSD) and System Development Environments (SDE), and discusses the order they should be deployed.

Chapter 18 discusses how and when to use Rapid Application Development, Prototyping, and Object Oriented development

Appendix A presents an example of estimating and planning for a project from general design onward.

Appendix B investigates the correlation between the estimating percentages and the work in each phase.

Appendix C defines the life cycle phases that are used throughout this book.

Appendix D lists the abbreviations used throughout the book.

OVERALL PRINCIPLES

There are two high level principles that will be presented in this chapter. They are:

- Project Phases Model, and
- Project Pyramid Model.

2.1   PROJECT PHASES MODEL

There are many phases in the life cycle of a project. Some of these are the General Design, Programming, and System Testing. For each phase in the lifecycle, there are three distinct stages. The stages are depicted in Figure 2.1 and consist of:

- Planning,
- Execution, and
- Review.

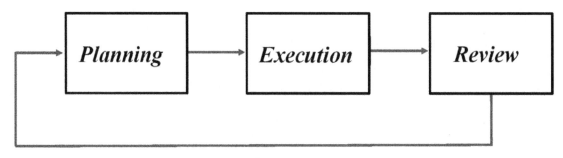

Experience

Figure 2.1

Planning

Planning includes all the activities involved in planning for the phase. This includes defining the scope, estimating the effort, and the construction of a plan.

Execution

This stage consists of executing the tasks that were defined in the plan to accomplish the phase. It is complete when the acceptance criteria have been accomplished. For many phases, the acceptance criteria consists of the approval of the deliverables from that phase. For the implementation, the acceptance criteria usually consists of the system passing the acceptance test.

Review

This stage involves a review of the prior phase. This does not refer to specific management and technical reviews that may take place throughout the project. Typically, formal reviews are usually done only after Implementation. On large projects, a review may take place after General Design[1], in order to provide feedback on the estimates and successes and failures of the General Design early. The review involves assessing the validity of the estimates, and recommendations to improve upon estimating, as well as general comments regarding what went well and what did not. It is important to review all projects and use the information to improve the process by applying the experience gained into future planning efforts.

Each of these stages represents important steps to the successful implementation of a project. For instance, if a project or phase is not planned properly and you begin the phase, problems will arise at an alarming rate. As phases and projects are completed, the experience that you have gained is used as input for future phases and projects.

2.2   PROJECT PYRAMID MODEL

There are a myriad of components and resources that constitute a project.

---

[1]See Appendix C for a description of the phases used throughout this document

Figure 2.2 depicts the components and resources that represent the *key success factors* in a project. They are arranged in a hierarchical or pyramid fashion. You require the building blocks from the lower levels to be successful, and you **will not** be successful unless **each** layer is in place successfully.

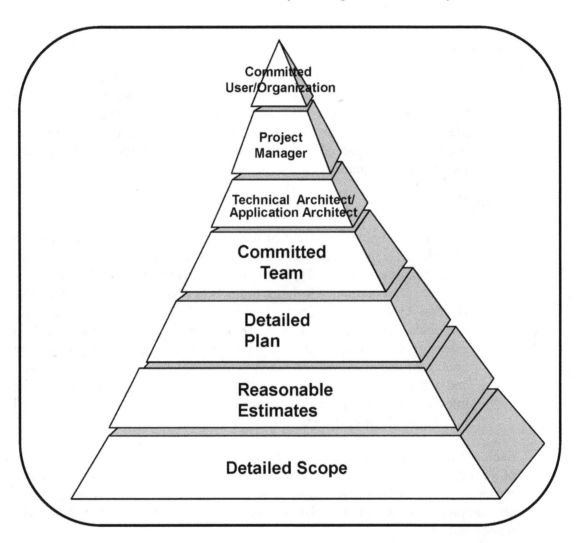

Figure 2.2

A major premise for this model is that it applies to projects that are fixed in terms of dollars (which correlate to man days) and time. You will be successful by ensuring that your project correctly addresses each of these areas. For time and material consulting, these factors are just as important, and can be applied in order to deliver a quality system as quickly and as inexpensive as possible, but, since there is no commitment to deliver the project at a certain time for a price, Organizations and Project Managers tend not apply the discipline and rigor needed to do so. It is, therefore, important to treat all projects as fixed, even if they are not, and apply the principles in this book to manage the project successfully.

Let us look at each of the pyramid components individually.

2.2.1   Detailed Scope

This component is at the base of the pyramid for a good reason. Without a well-defined project scope, most of the other components are unachievable. Without detailed scope, it is not possible to create reasonable estimates, and it follows, that the detailed plan will not be valid. Therefore, the Team, the Technical Architect and the Project Manager are trying to deliver something that is not defined, not estimated correctly and not planned correctly. The user organization may be extremely committed, but to what? They will not succeed. Chapter 4 discusses how to define the scope and objectives for a project.

2.2.2   Reasonable Estimates

Now that the scope is well defined, it is important that all tasks are defined and estimated. Based on the scope, the estimates have to be achievable and realistic, not just guessed at. The team must commit to the User that the phase or the project is going to be delivered for the man days that are estimated. If you view the project team as a separate company, you can visualize that if your estimates are low, you will overrun and lose money. If your estimates are greatly inflated, you may not "win" the business. For internal projects, this means that the cost to develop the system may not be cost justified in the business case and the project may not be funded. In the competitive systems integration business, what this usually means is that you do not "win" the contract.

You need to estimate the work accurately in order to "win" the business, and successfully deliver it. Chapter 5 and 7 discuss how to create reasonable estimates.

2.2.3   Detailed Plan

Once you have a well-defined scope and reasonable estimates, you need a good plan. Knowing what all the tasks are and the timeframes for each without creating a detailed plan will be like trying to go around the world, by car, without maps or assistance. Eventually you may make it, but it is not optimal. A detailed plan provides all members of the team and the Users with a complete guide as to which tasks and deliverables are their responsibility, how long they should take and when they are to be completed. It provides firm commitments to the costs and schedule and delineates the approach to managing the project. Chapter 6 discusses how to develop the high level plans while Chapter 8 discusses how to develop the detailed project plans.

2.2.4   Committed Capable Team

Yes, after a project, many team members may feel like they should be committed (to an asylum). This is not what I meant though. A good team is essential to any successful project. However, a team of superstars is not required, and can have a negative effect on the overall project, if their egos conflict. Having a good solid team is not enough though. They all should be committed to the following:

- the estimates and the plan,
- excellence,
- integrity, and
- the team.

Committed to the Estimates and the Plan

One could argue that in an ideal world, all tasks that are assigned to a person, should be estimated by that person. If they were, that person could commit to them easily. However, it is not always possible for everyone to estimate his own tasks, and even if it were, not everyone is a "good" estimator. What is important is that each person reviews the estimates and personally commit to them. In order to obtain this commitment the estimators need to have estimated properly. The estimators should be estimating the amount of effort for a task for a person at a certain level. Then this is what is expected of individuals at this level. If an individual is uncomfortable with the amount of man days allocated for a task, the task estimates should be reviewed and if valid, assigned to another person. Personal commitment is required.

Once the estimate is validated, the individual should commit to the Project Manager and the Team that he will try and accomplish the tasks assigned to him in the man days allocated.

Committed to Excellence

Everyone should **want** to come to work. Everyone should **want** to put his name on the work he has done and be proud of it. Everyone should strive to be the best he can be (sounds like a commercial for the US Army).

You can easily see the committed people are on your project or in your company. They are the ones who are going the "extra mile" to do it right. They are spending their own time to get things done, to help others. One person I know was so upset at being assigned to a new project before completion of the first, that she said to me "I'd work for nothing to see the system implemented", and she meant it.

Where do you find such people and how can you develop more of these people? Should you use money as an incentive to motivate people to do a better job? No! Reward excellent performance, but extrinsic reward should not be the main motivating factor.

Management's objective is to provide the proper environment to allow everyone to achieve his best. They must provide a working environment that allows people to get the job done, proper training to do the job expected, and clear career advancement possibilities. This is not an all-inclusive list, but indicates some of the management issues that need to be addressed in order for each person to be able to enjoy his work and grow. If employees enjoy what they are doing it is likely because they are accomplishing what they and management expect. They feel good about their accomplishments and management provides positive feedback as well. They will see that doing quality work really pays off, both from a career standpoint and from a personal standpoint.

Committed to Integrity

Each person is responsible for the quality of the work he does. In attempting to meet a schedule, a person may cut down on some of the quality, - for instance, during programming, he may not unit test programs as well as is necessary. This just causes problems in System Testing and Acceptance Testing. If an individual is behind, it is very important to be honest with the Project Manager and indicate how much he is behind. Keep in mind that all of the tasks are estimates, and it is possible that the reason for being behind is that the estimates were low.

If each person properly reports his true status, a Project Manager can take effective action. This may mean re-assigning tasks to individuals who are ahead of schedule, adding a resource, or waiting for a week to see if the individual can catch up. The point is that a proper assessment of the situation is possible. Everyone should be honest about his status.

Committed to the Team

It is important for all team members to know and believe that the project is a **team** effort. Everyone on the project, the **IS resources** - the programmer, the analyst, the Application Architect, the Technical Architect and the Project Manager, and the **User resources, is a team.** There is one goal that everyone is striving for, - successful implementation. Everyone needs to do what is required to ensure this. Everyone must be rowing together in the same direction. It is not an individual game. Help out your team members, try and get ahead of schedule so you can assist those who are behind schedule. **Everyone** succeeds if the project succeeds.

The other part of the statement included "capable". Just what do I mean that the team should be capable? Do all member of the team need PhD's. No. What I do mean is that they are truly capable of accomplishing the tasks that are or will be assigned to them. If they are a Junior Programmer, their tasks should be easier than the tasks assigned to a Senior Programmer.

Also, capable does mean at least average. It is unfair to the team, and the client to have a person that cannot accomplish tasks that are expected of them. For instance, on one particular project, a senior consultant (with more than 20 years' experience) was asked to create a System Test Plan. He was asked if he had done this before and he indicated he had. Even so, a draft table of contents was given to him and all sections explained. The first draft presented to the Project Manager was a disaster. It missed the entire point of the document. After advising the person of the problems, he set out to rectify them. The second iteration was only *slightly* improved. It became apparent that he was not capable of the task that at his level should have been second nature. The task was re-assigned to others who, due to the impeding deadline re-worked the document over a weekend and were able to present a draft of the document on schedule. This individual was not capable in the role he was providing to the team, he would have been capable if he was a junior resource who was assisting in the creation of the document instead of being the creator.

2.2.5   Application Architect

The Application Architect (AA) plays a vital role during the Project Definition phase, and to a lesser degree during the General Design phase. During the Project Definition, the new system is conceptualized. Before we start constructing the system, we must ensure that the system:

- meets the needs and expectations of the User, and

- re-engineers the process.

Meets the Needs and Expectations of the User

First and foremost, the system must address the real business objectives of the User and provide the functionality that is expected. In order to accomplish this, significant analysis of the current business process should be conducted. A common problem is that the analyst simply documents what the User is asking for without applying his knowledge to create a better solution. Often the system that is requested by the User looks like an automation of the current manual system. It meets the needs of the User, but likely it is not the best solution.

Re-engineers the Process

Re-engineering is a buzzword that when applied to a project means proper analysis of the process, where the process is composed of both the manual and automated components. By proper analysis, I mean that the analyst understands the current process, the requirements of the business and then, unencumbered by the existing process, conceptualizes how the new process should look. This is done for the whole system, both automated and manual, and takes into account the technology that is available to provide the best solution. By technology I do not mean the Enterprise Wide Architecture[2] that an organization may have in place, or is defining, but rather certain enabling technologies such as imaging or speech recognition which may be appropriate for a particular system and should be considered.

---

[2]See Chapter 17 for a discussion on Enterprise Wide Architectures

13

Another important area that the Application Architect is responsible for is ensuring that the current common processes are used, and developing new common processes. This will allow organizations to leverage an application base allowing for more rapid development.

Clearly, if the Application Architect has conceptualized a poor or invalid solution to the business needs, the rest of the project will not succeed in delivering a good system. At best, it can deliver on time and on budget a poor solution! This means that the Application Architect is playing a vital role and his ability to architect a good solution is a key success factor for the project as a whole.

2.2.6   Technical Architect

What is a Technical Architect, and why is he important to a project? The Technical Architect (TA), is responsible for understanding functional specifications and translating these into technology solutions.

On projects he should be assigned **full time**, and fulfill two important missions. First, he is responsible and accountable for delivering a solution that is integrated with the enterprise wide architecture. Second, he is responsible for ensuring the success of the project from a technical perspective. This responsibility includes:

- Ensuring a System Development Environment (SDE) is in place.

- Ensuring all hardware, software, and communications equipment is ordered, delivered, installed and tested when required by the project team.

- Ensuring design and programming standards are followed.

- Ensuring all technical problems are resolved in a timely fashion.

- Ensuring the quality of the system from a technical perspective.

- Ensuring Technical Objectives are met regarding such components as reliability, flexibility, maintainability, backup, recovery and performance.

- Maximizing the development productivity.

- Ensuring consistency of design and programming.

- Optimizing the operational costs.

While the Technical Architect is responsible for all technical components, including the overall design, he may also be called upon to design the data bases and program key components of the system. Also, although the Technical Architect is responsible for all of these components, he need not be an expert in all areas – but be able to understand the areas and call upon the expert resources as needed to complete the task at hand.

The Technical Architect is a **very** important role. The Project Manager should expect that the TA is ensuring that no technical problems are inhibiting the team's progress.

2.2.7 Project Manager

The Project Manager has a critical role on the team. No matter how well the previous responsibilities have been implemented, poor Project Management will dramatically increase the likelihood of failure, while good Project Management will dramatically increase the likelihood of success.

The Project Manager has a wide range of tasks and roles that span all facets of the project. He should be able to keep the scope contained, and keep all of the team busy and unencumbered by any politics of the project. In short, he should provide an environment in which the team can flourish.

The Project Manager has the toughest job on the project. It may seem that the Technical Architect or the Programmer who is coding his heart out has the toughest, but the Project Manager has the **responsibility** for successful delivery yet he can't speed the process up by doing some of the programming or design. What he must do are things such as ensure all people on his project are able to work at 100% of capacity, ensure no uncontrolled scope increases occur, adjust task order and people's tasks to optimize the project's chances for success. To be successful, the Project Manager must worry about the project every day. He must think about what could go wrong today, tomorrow, next week, next month, and take actions so they don't affect the team. It is a challenging and tough role.

The specific roles and characteristics of a good Project Manager are discussed in Chapter 16.

2.2.8 Committed User/Organization

Now we come to the final component on the pyramid - a committed user/organization. This is at the top for a reason. If everything else below it on the pyramid is perfect, but the user organization is not committed, then the project has little hope for success.

What do I mean by committed user/organization? First, there must be a champion of the new system, a Business Sponsor. Typically this is a senior individual that has responsibility for the new system. The Business Sponsor must want and need the system. It must be important that the project be completed as specified within the agreed cost and time frames so that the business benefits can be achieved. The Business Sponsor must agree to the scope and understand the relationship between increased scope and schedule and cost. He should be striving to obtain a system that meets the needs of the business and deferring changes to future releases unless they are critical.

The Business Sponsor must be able to drive his department and other departments to deliver what is expected of them. This includes making timely responses to decision requests, reviewing deliverables, and all other project responsibilities that are delineated in the Project Plan. This requires a strong individual especially when multiple departments are involved.

The organization as well must see the benefits of the system and support the undertaking, so that if the Business Sponsor leaves for some reason, the project will still be supported.

This all seems so intuitive and sensible, why would it ever not be the case? I don't have a good answer for that.

Some companies like Great West Life (GWL) Assurance Company have required that the Business Sponsor create and be the owner of the Business Case for a project. The Business Case clearly identifies the scope of the project, the benefits and the costs. The IS department is responsible for the IS costs of the project and the Business Sponsor is responsible for all business costs *and* for clearly delineating the benefits in *achievable* terms. If some functionality (some screens and reports) cannot be shown to deliver any benefits, they will not be included in the project. This has a dramatic affect. It should mean that only those externals that have some real *business* value are developed. This will result in smaller projects and ones that are better controlled since the Business Sponsor is responsible for ensuring that these benefits are achieved. There is a tight alliance between the costs and the benefits that will be achieved. If the costs for development increase, with no new functionality to increase benefits, the cost may exceed the indicated benefits. This clearly keeps the Business Sponsor and the business representatives focused on delivering what was in the

Business Case. As changes occur, and they always will, they are assessed in light of the costs and benefits. If there are real benefits for adding a new report (measurable ones) then the change will go through, otherwise it will be rejected, by the Business Sponsor.

Before the project is approved, GWL management reviews the Business Case. If they approve, there is the organizational buy-in to the project. After the project is completed, GWL management reviews the Business Case with the Business Sponsor to determine the accuracy of the benefit projections.

Here are a few examples where business sponsor and organization support have not been there:

> Components of a large justice project languished in Acceptance Testing for over two years. The components were statistical reports. Do you think management really needed them? No one really cared if they ever were completed, so the proper resources were not allocated and the proper discipline not enforced. (As it turns out, the System Acceptance Criteria was not done correctly either).

> A large real time fire dispatch system was in development for over four years and still had at least two more years left. It originally was scheduled to be completed in less than 3 years. What happened? The scope was not managed correctly and the system grew to over 3 times the original scope. As a result, rather than have an implemented system and be working on phase two, nothing had been implemented. If the User **really** needed the system as it was proposed, they would not have added so much functionality. Had the User been able to defer the non-mandatory changes to future releases, an already functionally rich system would have been up and running. They should have had the same goal as the development team, to implement the functionality as it was specified, as long as it was sufficient, and then implement a phase two of additional enhancements.

> A very large justice system was completed after 6 years of development. This included over three years in the General Design, and at least two technical platform changes. Now, the bad news. About 90% of the system was not implemented, it was shelved, and a package solution was investigated. How could this be? Clearly the vision and the goal of the "Business" were not embraced by the organization as a whole. It was deemed to be unsuitable. What a

waste of time, effort and money!

➤ A very very large hospital system was 3 times over budget and likely 3 years over its targeted implementation date (although no one really had a firm grasp on the exact dates). This project happened to be a time and materials contract. The vendor had a financial gain as long as the project continued. There were countless problems in this project (at least 4 for each pyramid component), but one of the most glaring was the lack of a single business sponsor. As changes arose, there were "consensus" meetings, or the de-facto sponsor would accept the changes without knowledge of how they would benefit the company. Finally, after 3 years, they appointed an individual that had the knowledge and authority to control this run-away train of a project.

3.0

PLANNING - OVERVIEW

No book, no course, no tool and no seminar will teach you how to develop a proper plan without you **thinking**. Each plan requires you and your team to think about the system you are developing and apply appropriate considerations in developing the plan. Some of the components that affect plans are:

- new technology,

- nature of project (Is it mainly a maintenance project?),

- number of Users,

- availability of Users,

- complexity of the system,

- number and complexity of interfaces, and

- risk.

The Project Manager must review his project in light of these details and craft a plan to accommodate them. No tool can know the intricacies and idiosyncrasies of your project.

Planning prepares a roadmap of where we are and where we need go. Without this roadmap we cannot tell where we are in relation to our goal and when it will be achieved. Every project but the simplest requires a well thought out plan. Watts Humphrey states "In any but the simplest projects, a plan must be developed to determine the best schedule and the anticipated resources required. In the absence of such an orderly plan, no commitment can be better than an educated guess"[3]

Let's review what the main objective of the Project Manager is:

---

[3]Watts S. Humphrey, *Managing the Software Process*, Addison-Wesley, 1989, p7

"To successfully deliver a project on time, on budget and to specification".

One of the keys to accomplishing this is:

"To develop an overall plan that has realistic estimates and is crafted using good project management principles".

What are "good project management principles"? This chapter will go through these principles in detail, enabling you to craft a good plan. The word craft is important here, because project management is a craft. It is not something that happens by simple entering the data into a project management tool. It takes creative thought

Figure 3.1

So how do you craft a good plan? Let me first define the "ideal" plan from a project manager's perspective. An "ideal" plan is one that is crafted using information required to construct the plan (i.e. number of interviews, or number of programs, etc). It is developed using a methodology that ensures that the right number of resources and the right skill levels are deployed at the right time in the project. This "ideal" plan is one that will be achieved easily, if all other project management principles are adhered to.

A "good" plan is one that applies a quantifiable amount of aggressiveness, and therefore a known amount of risk to an "ideal" plan. The aggressiveness is quantifiable, in that we understand the ideal resources and timeframe, and the amount we adjust from this is the risk. Thus if we modify the ideal plan to be accomplished in 10 months rather than 11 by adding a resource, we know that the plan is 1 month more aggressive, and by adding one more resource we have increased the complexity of communications and understanding by one person over the ideal.

The "ideal" plan can be tuned by doing such things as:

- adding more than the "optimal" resources,

- judiciously overlapping certain components, and

- starting some highly risky or dependent technical tasks early.

After this tuning, the plan is a "good" plan and one that the Project Manager should be able to accomplish (he understood the risks and judiciously altered his plan). Further reduction of the plan's duration to meet a "hoped for" date of the User will INCREASE the RISK and should ONLY be done if the User understands this risk and accepts the downside of not achieving the plan in terms of both dollars and duration. There are limits to even this approach. If the Project Manager believes that the plan is as aggressive as possible, and anything more will only lead to failure, he should stand his ground and state that the only way to accomplish the end date is to decrease the amount of functionality that will be in the first release. If this is unacceptable he should not accept the end date, but rather suggest again the futility of attempting to meet an end date that is not achievable. It is in no one's best interests to have a plan that suggests an end date, and then have the end date missed. By being firm on this point, it may seem like job suicide, but if you can articulate your reasons well, and there is reasonable management, they will likely appreciate your concern and diligence and take alternative positive action. If they don't, you may be replaced by a more "positive" Project Manager who will promise the delivery date you wouldn't, only to fail. Management should know the truth as early as possible, and not be led to believe that dates are achievable if they are not.

Both the ideal and the good plan should be crafted **without** being influenced by any predetermined end date. Any date the User may have given you should be

21

totally ignored until after the plan has been crafted. This is an extremely important point. For if you craft a plan with an end date in mind, you may add more resources than you should and create phase durations which are too short. The project is now high risk, or even unachievable. But since there is no baseline "ideal" plan or the more aggressive "good" plan to compare to, you are unaware of how aggressive or unachievable the plan really is. There may be no humanly possible way to succeed.

Figure 3.2 depicts the three types of plans, the "ideal", "good" and "silly" plans.

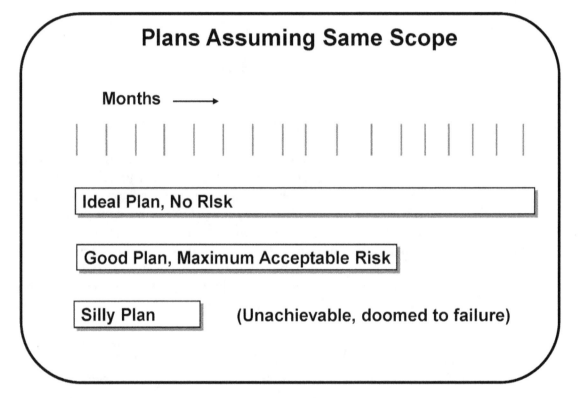

Plans Assuming Same Scope

Months ⟶

Ideal Plan, No Risk

Good Plan, Maximum Acceptable Risk

Silly Plan   (Unachievable, doomed to failure)

Figure 3.2

What has happened in the past is that the User has stated when they need the system by. The Project Manager then develops a plan to meet this pre-determined end date, disregarding project management principles. The Project Manager may not even know that the plans are unachievable. Meanwhile, the User expects delivery on time since *IS* created the plan. When the delivery date is not met, the User blames the Project Manager and the IS organization.

In today's world, Users constantly want a certain amount of functionality by a certain date. Why? How do they know if it is possible or not? Certainly business cycles demand that systems cannot be implemented during certain periods, or that systems in order to achieve the needed benefits must be implemented before a certain date. Examples of this are:

- A system to assist Registered Retirement Program processing must be up before February of a particular year.

- A retail point of sale system would not be implemented in December (the Christmas rush).

Clearly, there are legitimate reasons for requesting dates. However, the User must be flexible in the amount of scope that can be accommodated in that timeframe.

It is crucial to understand that the User can control the scope or end date, but not both. If they state that they need the system on a particular date, then it is incumbent on the Project Manager to work with them to see how much of the scope they can achieve. If they say they need a certain amount of the system (defined scope) then the Project Manager must craft a plan that will be aggressive, but achievable. The User can't say I need all of the scope and I need it by this date. Figure 3.3 depicts the relationship between the size of the system (scope) and end dates that are achievable.

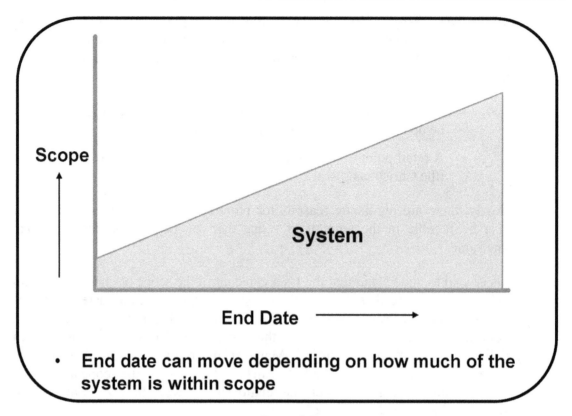

Figure 3.3

Certainly it is possible to deliver all of the functionality on the date requested, but this should be determined after the "good" plan is crafted, not before.

Recently a Request for Proposals (RFP) came out that requested some 120+ screens and 50+ reports to be developed. The General Design had not yet been completed. The system required a network, which had not been analyzed. They requested that the system be up and running (with trained users) in 4 months! Not likely. However, I did have trouble convincing the marketing representative that it could not be done. He thought we should create a plan that could accomplish their goals in 4 months. He asked "If it can't be done by any firm, why did they ask for this date?". Good question. Well, as mentioned before, if they could get the system in before the 4 months, there was benefit for that fiscal year. If not, the benefit would not be achieved for another 12 months. Therefore, they suggested it be done in four months. No one bid four months. However, if they would have allowed us to state how many of the screens and reports we could get done in the timeframe (control scope), something could have been up in four months (not a lot given that we still needed time to define the architecture in order to do a proper job.)

If the User has to have the system by a certain date, then the only recourse is *to reduce the scope of the project so that the effort can be achieved in the desirable timeframe.* This is an extremely important project management discipline, one that needs to be explained and then re-enforced with the end Users.

It is extremely important for IS and the Users to understand that the User can state a desired end date or the desired scope of the system, but not both. Given that the scope is defined, Information Systems (IS) can plan out the end date. Given the end date, IS can determine the amount of work that can be accomplished and working with the User, pare down the scope to the amount of work that can be accomplished in the timeframe.

An ideal project has a User and organization committed to the project and to the value to business of on time delivery.

PLANNING - SCOPE AND OBJECTIVES

It seems perfectly obvious that before you begin planning anything, you should define what you are trying to achieve (objectives), and how encompassing your objective is (scope).   Surely if you decide on painting your house, there is an objective - to improve the appearance, and you no doubt have decided on the scope - the kid's room, one bathroom, the living room, or the whole house, inside and outside. With those two decisions made, you can now plan on how much resource it will take, both people and paint, and plan out how long it will take, one hour, one weekend, or one month.

Figure 4.1

Ahhhh, it's so simple, isn't it. You wouldn't have just decided to paint the house for no reason, and then buy enough exterior and interior paint for the whole house, and **then** decide on what you were going to paint. Or you wouldn't have booked 4 weeks of vacation in order to paint, without deciding on how much work there was to do.

You wouldn't have tried planning and resource allocation until the scope and objectives were clear.

Whether in the Project Initiation Phase, the Project Definition Phase, or the General Design Phase, it is most important to define the scope and objectives and have concurrence between the team and the user as to what these are.

Often, it seems user departments will request from Information Systems that system initiatives be estimated without adequately defining the scope. Has anyone heard something like the following - "We need a new accounting system, let's get Information Systems to tell us the cost"? And worse, has anyone given them the answer, even if it was a ballpark? Ballparks do set expectations even with the caveat that it is a ballpark and can be out by 100 or 200 percent.

Before giving any estimate, the reasons for a new system should be documented. Depending on the phase you are in, different levels of detail of the strengths and weakness of the old system, and the new requirements would be included. These reasons for a system are given during business planning, both Strategic Planning and Tactical Planning and on projects during the Project Initiation Phase and the Project Definition Phase. For our purposes, generally, most estimates that the Project Managers will be asked to provide start at the Project Initiation or Project Definition phase. At this point, the objective of the new system and a detailed definition of the system's scope need to be clearly stated. In the above "accounting system" example, does the system include all aspects of AR, AP, GL, etc, or some component of that? Is it a companywide accounting system or only for cafeteria services? Does it interface with Payroll, Human Resources, and Marketing, or is it stand alone, etc.

The Project Manager having a clear understanding of the scope and objectives is not enough. Both the Project Manager and the User must have that same understanding. Many projects are doomed to failure the day after the contract is signed because there is not a good COMMON understanding of the scope. The User has one view and the team has another view of the proposed system (see Figure 4.2). This leads to User expectations that cannot be met within the timeframe and cost that were proposed. To avoid misunderstandings, define the scope in as much detail as possible, and in a specification that is understood by the Users. If the scope is not understood, but signed off, you are in as much trouble as not having a signoff at all.

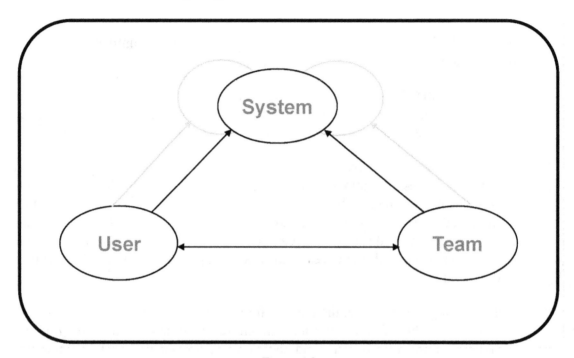

Figure 4.2

4.1   SCOPE AND OBJECTIVES FOR DIFFERENT PHASES

The objectives for a project remain basically the same during the course of a project, but the scope changes somewhat depending on the phase of the project. As a guideline, during the early phases, scope is defined by how much of the process the new system is to encompass, and after Project Definition, is more clearly defined by the externals, and their definition. The following sections indicate the level of the objectives and scope **prior** to initiating the phase.

4.1.1   Scope and Objectives for Project Definition

The objectives for a new system should be stated in business terms. The objectives will have been delineated during the Project Initiation phase.

The scope for Project Definition is which business processes are to be included in examining the problem or opportunity, and how many people will be interviewed to assess the current situation.

4.1.2 Scope and Objectives for General Design

Overall objectives for the project should be stated in business terms. These most often are the same as those from the Project Definition. They may be augmented if an innovative conceptual design uncovers new benefits that can be achieved, resulting in additional objectives, but the base objectives should not have changed.

The scope is the output from the Project Definition, which defines each process as well as a list and definitions of the expected externals.

4.1.3 Scope and Objectives for Detailed Design

The objectives are the same as those from General Design, unless a change request has been issued to include additional functionality, in which case the objectives are augmented.

The scope is the detailed functional specification from the General Design Phase, which includes a definition of each external, the logic specified, and for screens and reports, a prototype.

4.2 CLARIFYING SCOPE

What are the ways to define scope and avoid misunderstandings or differing views of the system?

- Clearly Document the Proposed System
- Visualize and Document Expected Man Machine Boundary
- Bound Unbounded Tasks
- Clarify Clients Requirements
- Avoid Certain Words in Proposal/Plan

4.2.1 Clearly Document the Proposed System

I'm surprised the number of times Users have said to me that they do not understand the system that they are going to get, but that they have signed off the General Design stating it meets their requirements!

What causes this to occur? There can be pressure from their management to get the system up as soon as possible and they do not want to delay the end of a phase while they come to a full understanding. This lack of understanding will only cause more problems, but they do not find this out until later.

Another cause is that the method to describe the User's requirements may be complicated, not explained, incomplete, prone to computer jargon, or some combination of these. Figure 4.3 depicts a diagram that is very confusing and hard to follow. With some modifications it could be much more readable.

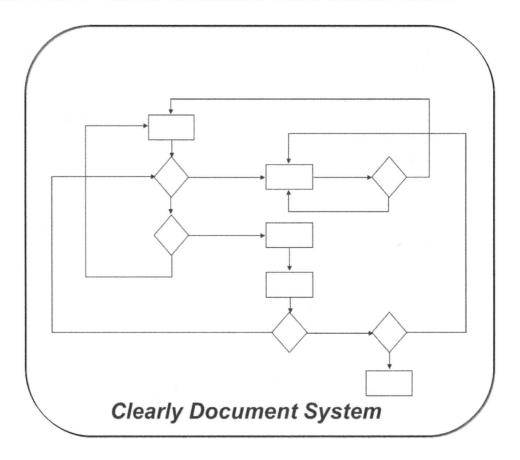

Figure 4.3

Clearly the intent over the years has been to describe the system in terms that the User and the Information Systems professional will understand. A combination of prototyping the system and the use of Object Oriented "use-cases" or traditional Data Flow Diagrams (DFDs), or Work Flow Diagrams (sometimes depicted using "swim lanes"), should be used to describe the new system (both automated and manual). The analysts should ensure that the constructs of use-cases, work flow diagrams or DFD's are understood. If they are, and the analyst knows how to use them, there should be no problems. However, if the analyst misuses the technique, it will confuse the User and the whole analysis effort will be jeopardized.

4.2.2 Visualize and Document Expected Man-Machine Boundary

Another effective way of meeting expectations is to ensure that the man machine boundary is known and well understood. Sometimes, even when good analysis tools are used, the man-machine boundary is not shown. This is a problem. The clearer the delineation is, the better the understanding of the system, and the clearer the end user's expectations are.

The systems analyst is supposed to provide a "conceptual new system", where "system" not only refers to the computer processes, but to the manual processes that interact with the **computer** processes. In analyzing and proposing the new system, it is possible to provide more or less automation, depending on the cost, the benefits and the maturity of the technical components. By clearly documenting the interaction from the manual to the computer system, the extent of automation should be clear, and no false expectations given.

4.2.3 Bound Unbounded Tasks

How do we bound unbounded tasks? Well first let's define what a bounded and unbounded task are.

Bounded Tasks

Bounded *tasks* are tasks that have a relationship between effort and task that is quantifiable, such as program X will take 5 days to program. The estimate is based on experience in the creation of similar programs. The program is **well defined** because the inputs, outputs and processing are defined. Given that these do not change, it is expected that this program will take 5 days.

Unbounded Tasks

An unbounded task is one in which the depth and/or breadth of the task is **not** well defined. The task itself may be clear, such as provide end user training for the new system. However, the effort to train the end users can take 1 day, 5 days, 100 days or 1000 days for the same system. Why? Because we may train 1 person or 100 people, we may have 1 all-encompassing 3 day course, or five 2 day courses, etc. Similarly, researching the applicability of a product to an organization's environment could take 5 days or 350 days, depending on the detail provided.

Figure 4.4

It is imperative that **every** unbounded task becomes a bounded task. This can be done in one of two ways.

Firstly you can make assumptions about the task that cause it to be bound, such as "we will train a total of 100 people, there will be 3 levels of courses, and there will be 5-8 people per course". An unbounded task is now bound by the assumptions made. If more people are required, or more classes, there are grounds to raise a **change request.**

Another way of bounding an unbounded task is to qualify what is expected, and then state the number of days that will be expended. For example, if the task is to review the applicability of a new software product to an organization's environment, you would want to determine how much effort should be spent overall, and then describe the deliverable by defining a Table of Contents (TOC) and estimating the pages of effort required. Given this, more detailed tasks should be defined that address one or more of the chapters, and the estimate for these should be commensurate with the degree of depth that is specified by the expected number of pages. Given that the TOC and the depth is understood by the team and the client, it is incumbent on the team member to accomplish the task in the given man days. If he exceeds the estimate it is probably because he is getting into more depth than the estimate provided for. It is up to the Project Manager to ensure that the tasks bounded by man days are accomplished by regularly ensuring that the correct level of detail is being addressed.

4.2.4   Clarify Clients Requirements

Often I have seen two people stating requirements entirely differently, but each nodding in full agreement with the other's explanation. The problem is that the spoken language is unclear as a specification tool. People can "read" into what a person is saying and determine that although stated slightly differently it means substantially the same as what they want. This sometimes happens because we think we are on the same wave length and given that we are close and agree with 95% of what a person says, we nod our heads. This causes many future problems.

In meetings where I have observed two people saying different things, but agreeing with one another, (and in some cases they are not even on the same wave length!), I interject and restate the requirement and ask for concurrence. I try to emphasize the difference between the two statements and even encourage one of the parties to disagree with my statement. When there is concurrence, I ensure that this is **documented** in no uncertain terms.

Another area that needs clarification is statements that define requirements for a system and include plurals of requirements. The most obvious example is reports. Many times specific reports are well defined, but then there is a catch all statement requesting the system to provide:

"all management and statistical reports"

This should sound alarms! What does this mean? This is another of those unbounded tasks. Request clarification of this immediately. Often the response to this is one of two:

"Oh those are our normal management and statistical reports that the current system provides."

If this is the case, ask for samples. Then include these as individual externals, ensuring that each is estimated properly.

Another example of unbounded reports is:

"We need management and statistical reports, but we haven't defined them yet"

Oh, good! You are supposed to commit to an estimate when the number of externals is not known and the complexity of them is not known! Avoid this like the plague. This needs to be bounded. There are two ways to bound something like this:

One way is to state an exact number that you will plan to, stating the complexity expected in terms that can be verified, such as "10 small reports, where small is stated to consist of reports that "have 5-10 data elements, are of a simple structure, and access 3 or less tables". Now, if an eleventh and twelve report are identified, they are clearly outside of the scope. Also, if a complex report is defined, it is subject to change control as well.

Another way to bound this is to state that you will provide an ad-hoc query tool that allows end users to formulate their own simple reports. Your obligation is then to include the tool in your costs (if it is not already at the data center), the costs to research the tool, and the training for the tool (remember to bound this too).

4.2.5 Avoid Certain Words in Proposal/Plan

Certain words will lead to expectations that are not valid. If you see these words in the Proposal, in the Project Objectives Document (an output from the Project Initiation Phase) or they are used during discussions of scope, you must clarify the point.

Here is a list of the 7 words that should not be allowed in documenting or discussing scope:

- *ALL*
- *EVERY*
- *EVERYTHING*
- *EVERYONE*
- *ANY*
- *NEVER*
- *GUARANTEE*

Each of these words should ring a bell, set off a siren in your head. Examples where they have been used include:

"Train **every** user"

"**Guarantee** the system is **never** down"

"Produce **all** management reports"

These words will cause problems if the real scope of the statement is not determined and agreed to.

5.0

PLANNING - ESTIMATING

OVERALL PROJECT

Users request estimates at different times in the life cycle, and we can provide estimates for some and not for others. Typically the requests are at the end of the following phases - Project Initiation, Project Definition and General Design. Let's first review what our general response should be to a request for the "overall project costs and duration".

Project Initiation

No estimate should be given. An estimate implies some rational thought based on facts has taken place. However, at the end of Project Initiation, we have very little to work with. The system's scope is not clear enough to provide an estimate for the overall project. We can and must provide an estimate for the next phase, but we must resist giving an estimate for the overall project.

If the User needs this for planning purposes, then the best we can do is request from them how much the anticipated annual benefit of the proposed system is, and then suggest that one goal of the project should be that the costs to develop the system be less than twice the annualized benefits. It is important than the User understand that this may mean a constant "reigning in" of the scope throughout the Project Definition phase. The User must not only agree to this, but actively support and acknowledge this to any User participating in the Project Definition phase.

Project Definition

After Project Definition, we should have a complete list of all externals, and a brief description of each. At this point we can estimate each of the externals (from a programming perspective) and develop an overall estimate for the project. Since we do not have the full and detailed specifications for each external, our estimate is not as accurate as after General Design. When providing the estimates to the User at the end of this phase, it is important to qualify the degree of certainty of these estimates. In general, the estimate provided should be within plus or minus 25%. In most instances the estimate does seem to gravitate to the upper end. In any material provided, this caveat should be included, so that the User understands the estimate is not "carved in stone".

General Design

After General Design, we should have a complete specification for all programs. All edits and logic, all screen and report layouts should be finalized. There should be no questions about the functionality of any part of the system. Given that this is the case, we can provide an estimate for the whole project that we believe is correct and that we can be held accountable for. We should be able to fix the costs and duration for the entire project, and deliver to these.

The estimating methodology presented in this section is valid for any development platform - mainframe, client server, internet, 2 tier, 3 tier, etc. The difference lies mainly in the guidelines for the programming effort. A simple mainframe report (yes they are still developed) program will take more time than a simple report program using one of the many report generation tools like *Crystal Reports*. So the guidelines must be tuned. In addition, you must take into account the existence and robustness of a Systems Development Environment (SDE), the availability of reusable code, and the degree of new development versus maintenance of existing programs. Each of these will affect the amount of time you estimate for a particular program.

This section discusses how to create an overall estimate given a specification that has defined the externals. An external is any report, screen or other process (batch, common, communication, etc) that needs to be **programmed**. For specifications that are not at this stage, **NO** overall project should be estimated, but rather the Project Definition (or General Design) phases should be estimated. Guidelines for estimating the Project Definition phase are presented in Section 7.

Throughout the estimating and planning exercise, a list of assumptions should be started and updated. Wherever and whenever you assume something, write it down!

5.1   TYPES OF ESTIMATES

There are a variety of estimating rules, tools, methodologies, and hand waving that is used to determine the size of a system. Unfortunately, typically, there is not a great amount of discipline and effort when using these. Often a lack of familiarity with the tool/methodology, etc, results in a **guesstimate, rather than a meaningful estimate**. To add to this problem, after projects are completed, project audits are generally not done and there is no comparison of **estimates** to **actuals**. This lack of comparative data prevents most organizations from improving the accuracy of estimates over time. Nevertheless, it is possible to conduct a reasonable estimate even if there is little history to compare to. There are three common methods used to construct an overall estimate: the Program Effort, the Lines of Code (LOC) and the Function Point method. All three methods can be used to quantify the programming effort for an external. This document uses the Program Effort method as an example, but does not imply that it is superior or should be used instead of the other methods. The methodology described in this chapter is applicable regardless of the estimating method employed.

As with all estimates, at **least** two and usually three people should prepare the estimates. If only one person does the estimate, there is no way to compare and discuss alternative views on the complexity of any external. Moreover it is **imperative** that the estimates are done independently of one another. A group consensus on an external is not the way to obtain the initial estimates. Don't worry that your estimates are going to be different from others and therefore people will say you are a "bad" estimator. I've had individuals who consistently estimate twice the amount of effort required (backed up by actuals) and others who are consistently 1/2 to 1/4 low (again backed up by actuals). However, after these tendencies are known, they provide a good estimate platform. At the outset, they do cause problems, and if you have large variances amongst a set of inexperienced estimators, you may want to get another estimate from an experienced estimator. For the purposes of this document we will assume 2 people are doing the estimate, person X and person Y.

Figure 5.1

Figure 5.2

Again, it is important to emphasize that the scope must be well defined and the previous phase completed. If all externals are not known, a reasonable estimate cannot be made.

5.2   PROGRAM EFFORT ESTIMATE

This method uses "rule of thumb" for categorizing different externals into different sizes. Prior to using this method, all externals need to be defined.

The procedure to be followed is:

1)   Create spreadsheet of externals

2)   Estimate all externals

3)   Combine estimates

4)   Synthesize estimates

5)   Review and resolve inconsistent estimates

5.2.1 Create Spreadsheet of Externals

Using a spreadsheet tool such as Microsoft's EXCEL, create a spreadsheet that has all the externals on the left side and 3 columns on the right side: one for Person X's estimate, one for Person Y's estimate and one for the "USED" estimate. The "USED" column is for the estimate that we decide is the best one to proceed with. It is best that the externals are numbered and the descriptions easy to understand, so the estimator will be sure to know exactly what external he is estimating, and so all of the estimators are estimating the same externals. Additionally, it is useful to partition the estimates by category; for example, the categories of SCREENS, REPORTS, and OTHER. Totals should be summed for these categories and then a grand total provided. If there are different environments, such as some programming on a server and some on a workstation, it is useful to break these out and total them separately. Spreadsheet 5.1 is an example of such a form.

Sample of estimates form				
Nbr		PERSON	PERSON	"USED"
		X	Y	
	SCREENS			
S1	add client			
S2	add insurance			
S3	add mutuals			
S4	calculate annuity			
S5	change address			
S6	change deposit amts			
S5	Menu			
	Total Screens	0	0	0
	REPORTS			
R1	clients by area			
R2	revenue by client			
R3	revenue by type			
R4	commissions by agent			
	Total Reports	0	0	0
	OTHER PROCESSES			
P1	calculate dividends			
P2	calculate interest			
	Total Other	**0**	**0**	**0**
	GRAND TOTAL	0	0	0

Spreadsheet 5.1

5.2.2 Estimate all Externals

Each estimator is then given the spreadsheet and the document in which these externals are documented. Prior to reviewing each of the externals and estimating them, a common agreement on the definition of the size of an external and the man day estimate for that size must be determined, understood by everyone and agreed upon.

In utilizing the "program effort" estimate, it is important to utilize the man days

41

that are applicable to the technology. For instance, a medium program written in MicroFocus COBOL may be 10 days effort, but the same program written in a higher level language or generator such as C++ may be only 5 days effort. The estimates that follow are based on a MicroFocus COBOL type language. When using a different language, you must alter these estimates accordingly. Different tools affect these estimates to different degrees.

In general, the high level languages such as Visual Basic and C++ will allow the "very small" and the "small" to be done much more quickly, the "medium" to be done somewhat more quickly, and the "large" to be generally unaffected. This is because most of these languages and tools alleviate or make easier the tasks of screen handling, error handling, editing and data access. These components are up to 80% of a typical very small or small program. When a program has more logic, which is true for medium and large programs, these languages do not provide the same degree of benefit, and in the case of very large (complex) programs, they may not provide any benefit. In fact, for these complex logic programs, often an exit to a more powerful language is used to accomplish this logic.

How do you adjust the man day estimates for a particular language? There are a few ways of determining this:

1) Review historical estimates of tasks that used the language.

2) Review vendor statements (Caution -Do not **use** these without some other guideline - they are typically **very** aggressive and although not wrong per se, they are obviously geared to a "perfect" environment with typically very simple transactions, and very experienced developers).

3) Consult with in-house experts or those who have used the language before.

4) Consult with outside resources for guidance and counseling.

The more thought and care that is used in developing the guidelines, the more accurate the estimate will be, which results in a more accurate and reasonable and **good** plan.

One further note of caution. When utilizing a technology or language for the first time, bear in mind the learning curve. It will likely take a month or two to become productive in the new language. Therefore, on small projects that use a technology for the first time, do not be aggressive, be conservative.

Now, back to the estimating process. For each external the estimator determines the **program** effort to be one of the following: **very small, small, medium, or large**, and allocates the appropriate man day estimate to that external. The following estimates are valid for the programming (including unit testing) using MicroFocus COBOL programs that have the described characteristics:

Screens

Very Small	-	less than 5 attributes
	-	only simple field level edits
	-	1 or fewer tables accessed/updated
	-	typical ranges are 2-3 man days
	-	as a standard use **2** man days
Small	-	6-9 attributes
	-	mostly simple field level edits
	-	simple logic
	-	2 or fewer tables accessed/updated
	-	typical ranges are 4-6 man days
	-	as a standard use **5** man days
Medium	-	10-15 attributes
	-	medium amount and complexity of logic
	-	some screen manipulations (scrolling, etc)
	-	fewer than 4 tables accessed/updated
	-	typical ranges are 7-11 man days

| | - | as a standard use **10** man days |

Large	-	10 or more attributes
	-	complex logic
	-	sophisticated screen manipulation
	-	greater than 3 tables accessed/updated
	-	anyone of the above or combinations could result in a "large" program
	-	typical ranges are 12-?? man days
	-	as a standard use **15** man days

Note: Although the "large" category could be used for any size of large program by adjusting the value upward, one must review any screen that takes in excess of 15 man days to program and question the design. Generally, it will be easier to break a complex screen into two or more smaller screens that are easier to program and easier for the User to learn.

Reports

Very Small	-	less than 7 attributes
	-	simple structure (all data easily fits on report)
	-	accesses 2 or fewer tables
	-	can be easily generated by a tool
	-	as a standard, use **1** man day
Small	-	5-10 attributes
	-	simple structure
	-	accesses 3 or fewer tables

	-	as a standard, use **4** man days
Medium	-	10 or greater attributes
	-	medium structure (all data fits on report, but there are some format changes, such as group totals, or not printing some fields unless they change from the line above)
	-	accesses 3 or more tables
	-	as a standard, use **8** man days
Large	-	10 or greater attributes
	-	complex logic or structure in order to present the data, such as group and sub-group totals, page breaks, etc
	-	accesses 4 or more tables
	-	as a standard use **15** man days

Note: In general, the number of reports should be kept to a minimum. In the Project Definition or General Design phase, always ask why an on-line screen couldn't be used (in most cases it could).

For screen and report estimates, use good judgment when determining which category it should fall into, **do not** blindly use these guidelines.

If you think strongly that it is "in between" a category, use the value you feel comfortable with. For instance if you think a screen is not quite a medium but more than a simple, use 7 or 8 person days. However, be prepared to defend your selection.

Other

The other category is for programs that need to be written that are not screens or reports. They include such things as download programs, upload programs, file synchronization programs, common routines, etc.

By their very nature that they are not well defined, they cannot be categorized with estimates for very small, small, etc. Experience is the only mechanism that can be employed.

Rather than try and estimate a complex program, break it down into the functions that it must do, and then estimate the smaller components. For instance, you may need an upload program to upload data captured on the remote Laptop to update a server. What exactly is required?

- a SEND program on the Laptop,
- a RECEIVE program on the Server, and
- an UPDATE program on the Server.

Now you have three programs you could estimate. But let us look at the problem further.

For the SEND program, you need:

- the ability to gather either 1 transaction at a time, or multiple transactions and queue them,
- a reformater to reformat the data for the server, and
- a communication requester to begin sending and to acknowledge proper receipt at the server.

Now you have even more detailed information with which to estimate these components. Similar breaking down of the RECEIVE and UPDATE programs allow for proper "thinking out" what is **really required** and better estimates because you know more about what really is to be done in the "UPLOAD" program.

5.2.3 Combine Estimates

Gather each person's estimates. Put all of the estimates together on the "master" spreadsheet. If you have given each person his own copy of the spreadsheet and asked him to update the column under his name, this will be much easier. If not, you will have to enter the data. Spreadsheet 5.2 shows an example of this.

Sample of estimates				
		PERSON	**PERSON**	**"USED"**
Nbr		**X**	**Y**	
	SCREENS			
S1	add client	15	15	
S2	add insurance	10	15	
S3	add mutuals	10	15	
S4	calculate annuity	15	25	
S5	change address	10	10	
S6	change deposit amts	4	5	
S5	Menu	2	2	
	Total Screens	**66**	**87**	**0**
	REPORTS			
R1	clients by area	8	10	
R2	revenue by client	12	8	
R3	revenue by type	8	10	
R4	commissions by agent	12	12	
	Total Reports	**40**	**40**	**0**
	OTHER PROCESSES			
P1	calculate dividends	30	25	
P2	calculate interest	15	20	
	Total Other	**45**	**45**	**0**
	GRAND TOTAL	**151**	**172**	**0**

Spreadsheet 5.2

5.2.4 Synthesize Estimates

Now that you have the estimates for each person, you need to go to each external and observe the difference in estimates. For estimates that are exactly the same, place the value in the "USED" column. When there is a difference of 20% for estimates up to 10 days, select the one you think is the best. Thus if one estimate were 10 days and another 11 use either 10 or 11, whatever you think is the best, and place this in the "USED" column. For estimates greater than 10 days, the percentage variance should decrease to 10%. Thus an estimate of 18 or 20 could go either way (10% of 20 days = 2 days). You may want to alter between the high and the low as you go along. For all estimates that are resolved by the above review, place the estimate in the "USED" column. For all others, the "USED" column should be left blank. Spreadsheet 5.3 shows our example after the synthesis.

Sample of synthesized estimates				
Nbr		PERSON X	PERSON Y	"USED"
	SCREENS			
S1	add client	15	15	15
S2	add insurance	10	15	
S3	add mutuals	10	15	
S4	calculate annuity	15	25	
S5	change address	10	10	10
S6	change deposit amts	4	5	5
S5	Menu	2	2	2
	Total Screens	66	87	32
	REPORTS			
R1	clients by area	8	10	8
R2	revenue by client	12	8	
R3	revenue by type	8	10	10
R4	commissions by agent	12	12	12
	Total Reports	40	40	30
	OTHER PROCESSES			
P1	calculate dividends	30	25	
P2	calculate interest	15	20	
	Total Other	45	45	0
	GRAND TOTAL	151	172	62

Spreadsheet 5.3

5.2.5 Review and Resolve Inconsistent Estimates

First, ensure that each estimator has a copy (hard copy is best) of the combined estimates done in the above step. Now, for each external that does not have a value in the "USED" column, discuss the reasons why each person estimated as they did, and determine the "right" value. A reason for the discrepancy could be that one person knows more about the process and therefore is more accurate. However, the "right" value can be the low, the high, an in-between value, or quite possibly a lower or higher value. How can it be lower or higher than the original estimates? Discussions of variance often leads to a more complete understanding and hence a new estimate. As each external is resolved, place the estimate in the "USED" column.

One final review should be done at this point. For all of the externals that are "large" a brief discussion of each should take place to ensure that all of the estimators agree with the function and complexity of the external. Sometimes different estimators will estimate the same external as "large" for different reasons. By discussing each of these it may be apparent that the external is larger than all of the estimators had thought and needs to be re-estimated.

After all externals have been resolved, you now have a reasonable programming estimate.   Spreadsheet 5.4 is our example, updated after the review and resolution of our estimates.

	Sample of final "used" estimates			
		PERSON	**PERSON**	**"USED"**
Nbr		**X**	**Y**	
	SCREENS			
S1	add client	15	15	15
S2	add insurance	10	15	10
S3	add mutuals	10	15	12
S4	calculate annuity	15	25	18
S5	change address	10	10	10
S6	change deposit amts	4	5	5
S5	Menu	2	2	2
	Total Screens	**66**	**87**	**72**
	REPORTS			
R1	clients by area	8	10	8
R2	revenue by client	12	8	12
R3	revenue by type	8	10	10
R4	commissions by agent	12	12	12
	Total Reports	**40**	**40**	**42**
	OTHER PROCESSES			
P1	calculate dividends	30	25	25
P2	calculate interest	15	20	20
	Total Other	**45**	**45**	**45**
	GRAND TOTAL	**151**	**172**	**159**

Spreadsheet 5.4

6.0

PLANNING - DEVELOP

HIGH LEVEL PLAN

From the estimates that were developed using one of the methods described in the previous chapter, you have an estimate for the overall programming effort. You now want to use this programming estimate to get overall estimates for all phases in the life cycle. Then, you can plan the overall phase durations using a Gantt chart. Finally, using the high level Gantt chart and the overall estimates, you want to map out each person's involvement in each phase on the low level Gantt to ensure you have the correct number of resources and have not underutilized or over utilized any one resource. Development of the high level plan is accomplished by doing the following:

1) Generate man days for all phases

2) Review deliverables required

3) Create high level Gantt (phase level)

4) Tune the high level Gantt

5) Add User required time

6) Create work plan for first phase

7) Create low level Gantt (phase level)

8) Tune the low level Gantt

If you have diligently followed the prescribed approach, the high level plan should provide a man day estimate and duration that you could confidently commit to.

6.1 GENERATE MAN DAYS FOR ALL PHASES

From the programming man day estimate, you want to obtain estimates for the standard set of phases. But, before converting the programming effort into an overall estimate, review the "other" programs. Do all of these programs require the utilization of all phases? Often they do not. For instance, a communication module used to pass data from a Laptop to a Server certainly needs to be analyzed, designed, programmed and system tested. However, it does not require User Training and will not require documentation in the User Manual. It should be excluded when converting to an overall estimate, and then re-entered into the phases that are required. For a more complete discussion of this, see the example in Appendix A. In the example on spreadsheet 6.1, it is assumed that all 159 programming days require the same consideration. To obtain the estimates for each phase, simply take the programming man day estimate (159 days) and multiply those days by the percentage associated with each phase. This then gives the overall project effort. See Appendix B for a full discussion on why these percentages have any correlation to the work in each phase.

Conversion of 159 Programming days to overall Project Days		
PHASE	**Percent**	**Man days**
General Design	22	35
Detailed Design	44	70
Programming	100	159
System Test	34	54
Acceptance Test	22	35
Manuals	16	25
Training	16	25
Project Management	34	54
Technical Architect	20	32
Conversion	0	0
Facilities Install	0	0
Total		**490**

Spreadsheet 6.1

Now, review each of the phases and decide which ones are required for this project. You may have completed General Design or there may be no conversion and therefore these estimates would go to zero. Also, at this point we are solely concentrating on the IS time, not the User required time. As such, some of the major functions may be heavily concentrated in the User's time and not IS. In particular, often times the user department is tasked with the Training and development of the User Manuals. If this is the case, there still will be some IS time required, but not as much as if IS was conducting the tasks themselves.

Training

Typically the Training component is concerned with only the end user training and not the training for maintenance personnel. If there is a specific reason to train the IS people, make sure you have allocated appropriate time to do this. To determine the effort, write down the specific components of the system that the maintenance people require, and determine the effort to teach them these components.

For the end user training that is conducted by the Users, we must ask ourselves "What will be required by the User from Information Systems (IS) in order to accomplish the training"? Clearly they need some assistance. They will need someone familiar with the system to answer questions and provide the trainers with some training. Someone will need to set up a training environment for them and document how the User can manage it (Specifically, how do they reset the system back to its original data contents at the beginning of each training session?).

Make sure there is documented agreement (an assumption in the plan) on the number of people to be trained, the timeframe, and the fact that the Users are conducting this and not IS.

Manuals

Manuals are composed of two components - operational documentation and end user manuals. For the operational documentation, we need to determine what documentation is required by production control and estimate the effort to produce it. With the end user manuals, if these are to be produced by the User, someone knowledgeable will be required to provide answers and guidance. If there is a group of people whose primary job is to write user manuals, discuss with them how much time is required. This will be a factor of how many externals there are, how complex the system is and how different the look and feel is compared to other systems the users are familiar with.

Now, back to the overall estimates. Sometimes, the phases will take more effort than the percentages suggest. There could be particular technical challenges or risks such as using a technology unfamiliar to the organization, or the need for a new Systems Development Environment (SDE). If so, you would want to do a high level task estimate. This forces you to think about what extra is involved and to document the effort. This time is then **added** to the Technical Architect (TA) time. Not all time allocated to the Technical Architect will necessarily be done by the TA; he may delegate some work such as development of some of the SDE.

Finally, we need to look at the last two phases.

Conversion

> Conversion can be anywhere from a few simple screens for data entry, to a sub-project larger than the system itself. List all files that need converting, all tables, and all manual data. Convert this effort into externals required. For example, will the User enter data manually from an on-line screen or into a flat file for batch conversion? From the externals, estimate the programming effort, and convert this into an overall estimate. Many of the "normal" phases such as Manuals and Training may not be required. For small conversions, the Project Management component will be done by the overall Project Manager and therefore adds zero to the project totals. However, for large conversions there will be a Team Lead that will require Project Management time. Add the total conversion time to the overall spreadsheet.

Facilities Install

> What are the hardware, software, communications and facility requirements for both the development and the production system? Many times there exists an infrastructure and all that is required is adding the new software to their workstations. Nevertheless, this must be done and an estimate is required. In other instances, hardware and software for workstations and LANs may have to be acquired (or upgraded), installed and tested for development and again on a larger scale for production. For this phase, list all of the tasks required. It will be of great benefit to have technical resources familiar with your target environment help in preparing the task list and estimating the effort and

skill level required. Add these estimates to the overall spreadsheet.

At some organizations, if the project requires a great deal of work to install facilities, this will be handled by a separate project team with dedicated responsibility to acquiring, installing, and testing the facilities. Given this direction, it is still important to determine what is required. If it is a large undertaking, then it will be separated out. If not, the tasks and deliverables associated with the effort need to be managed by the Project Manager in this Project Plan. Again, because the effort is not going to be large, the tasks and deliverables should be planned and managed under the phase where they occur. For instance if some workstations are required in the middle of General Design and more at the start of System Testing, then the plans for General Design and Programming should include the tasks, deliverables and schedule for their completion.

Now you have an overall estimate of the effort. However, it is still an estimate, and more work is required before you will want to commit to it. This synthesized version of the estimate is depicted on the spreadsheet 6.2.

Synthesized version of the overall project total			
	Original	New	User
	Project	Project	Required
PHASE	Total	Total	Days
General Design	35	35	
Detailed Design	70	70	
Programming	159	160	
System Test	54	55	
Acceptance Test	35	35	
Manuals	25	5	20
Training	25	5	20
Project Management	54	55	
Technical Architect	32	30	
Conversion	0	0	
Facilities Install	0	0	
Total	490	450	40

Spreadsheet 6.2

55

Note that we have added a column for the User Required Days. We did this because we had an estimate for the total effort, and had decided that the tasks in these phases would primarily be done by the User, and we did not want to lose our initial estimate of the task (here we have divided the 25 days into 5 days for IS and 20 days for the User). In a future step, we will validate the days for these tasks as well as add more of the User's time to other phases.

6.2   REVIEW DELIVERABLES REQUIRED

At this point in the planning exercise, we want to review the overall scope of the project from a deliverable perspective. Every organization ought to have some methodology that describes all of the tasks and deliverables in each phase. It is assumed that such a methodology exists that describes each deliverable, who creates it, who reviews it and who signs off the deliverable. Depending on the type of project, the technology, the phases completed, etc, some of the deliverables that are in the methodology may be eliminated, down-scoped or internally created (not subject to user review). In addition, some of the deliverables are more appropriate to a task estimate, since something like a "vertical prototype" will take a different amount of effort depending on the scope, technology, language and available tools. The impact of these decisions will result in adjustments to our overall plan.

The documenting of deliverables by phase will be required later, when we create the detailed project plan, so it is best to do it on a common standard word processor.

Essentially what you will document is a table with the Phase and Deliverable on the left side, and the delivery date on the right. For now, we need not concern ourselves with the delivery date (usually expressed in man months from start). What we do want is to list by phase all deliverables that are required for this project. We don't get bogged down in these details now, but quickly ascertain if we need the deliverable, and if so if its scope is as stated. An example of the documentation style follows:

PHASE	DELIVERABLE	DELIVERY DATE
General Design		
	Horizontal Prototype	
	Functional Specifications	
	...	
	Project Plan for Detailed Design	
Detailed Design		
	Technical Specifications	
	Program Specifications	

Continue this review for all deliverables in your methodology. If all deliverables are required and are as stated, we need not adjust the estimates. If there are only minor adjustments expected, they need not affect the estimate. However, if there are a number of deliverables not required for this project, the estimates for that phase where they *would have* been created should be decreased.

6.3   CREATE HIGH LEVEL GANTT (PHASE LEVEL)

Now comes the real fun part. How do you take the estimates, and decide how many people are required and for how long? An excellent book on the effect of adding resources to a project, is Fred Brooks "The Mythical Man-Month: An Essay on Software Engineering"[4] which details how adding more people to a project reduces the output. In real life I have seen a project double in size over 8 months (from 20 to 40 programmers), only to be further behind after the 8 months! Why? - lack of communication with all team members, too many people to "go through" to find the answer, easier to do it yourself than to find out if and where a common routine exists, etc. Essentially, the reason is lack of strong communications due to the size of the team.

There is an optimal number of people to assign to a task for an ideal amount of

---

[4]Fredrick P. Brooks, *The Mythical Man-Month: An Essay on Software Engineering*, Addison-Wessly, New York, 1975

time.   In creating your plan, you should use the right number of people for the right number of weeks/months in each phase, and each phase should have the right duration. Oh, that clarifies a lot!

I'll get to the definition of "right" in a moment, but first I want to emphasize that the plan you are crafting is the "ideal" one. That is, if you had unlimited time, budget and resources, the number of people and the phase durations you have configured would be used. This essentially is a plan with little risk (if your estimates are accurate). Often, however, you will be asked to get it done more quickly. You can do certain things to decrease the duration, such as adding people, or overlapping phases, but you are **increasing the risk** that the project will not end when you have planned. This is fine, as long as you, your management and the user management **know** and **understand** that you are being aggressive. This results in the creation of the "good" plan. If you do not understand the risks of adding people and overlapping phases, you may add too many people, which will result in lost productivity (or worse), or you may overlap phases too much and be unable to accomplish tasks in either phase due to dependencies. For instance, you can overlap some system testing with development, but you can't have them totally overlapped. Moreover, the amount of overlap is dependent on the phase duration - on a 1 month, 1 programmer, 1 program programming phase; you cannot overlap system testing at all. But, with a 10 month 12 person programming effort, you can overlap some execution of System Testing (if you must).

Again, I should emphasize that once you do take the ideal plan and add additional risk, you should do so without trying to achieve a given end date. You can add people, overlap phases as much as possible (without adding too much risk), and if the end date is still beyond the User's expectations, the only option left is for the User to *reduce scope*. If you arbitrarily add people and overlap resources without understanding the risk and the chances for success, you will most likely fail.

The next step is to translate the man days for a phase onto the Gantt chart. At this point there is a tendency to enter data into a planning tool and out will pop the answer. Resist this temptation. It is best to manually plan this and the low level Gantt, and **then** enter the plan into an automated tool (you will know the duration and man days before entering the data). There are several ways of doing this translation. The basic goal is the same though: to use as *few* people as possible to accomplish the phase in a reasonable timeframe.

An experienced person would be able to look at the man days and decide on the optimal number of people for the optimal number of months, based on having done this so many times in the past. See Appendix A, for an explanation of the related thought process of an experienced project manager. There is, however, a fairly standard mechanism, called the square root method that provides you with a ballpark duration and resourcing load fairly quickly.

Basically, by taking the square root of the number of man months for a phase, you get the duration and the number of people to be assigned. Caution and common sense both must be exercised. The algorithm works well for non-overlapped and fully loaded phases such as General Design, Detailed Design, and Programming. It requires some tuning for System and Acceptance Testing. These two phases are generally overlapped with the preceding phase, and are staffed with fewer resources during the overlapped time. Keeping this in mind the duration provided by this method is useful in determining the overall duration's reasonability.

Let's look at an example that shows how the data can be used to craft the plan. For the purposes of planning, use 20 days per month as an average number of work days. At the high level, add 5 days to the Gantt about every six months to accommodate statutory holidays. This is usually added to the end of the major phase that is completed near each six month timeframe. These two steps provide for 20 days over a whole year for vacation, sickness, and other non-project specific work.

Let's look at a General Design phase estimate of 400 man days. The algorithm says take the square root of the man months for the phase. So we take 400 man days divided by 20 days per month, which equals 20 man months. The square root of 20 is 4.472. This is the suggested duration in months, and the suggested resource load. Clearly, we need to convert these numbers into something more realistic - closer to whole people for whole weeks. If we do this, we have either 4 people for 5 months or 5 people for 4 months. .

Continue applying this method for the Detailed Design, Programming, System Test and Acceptance Test phases. With the System Test and Acceptance Test, the estimated durations are fairly reasonable when overlapped.

System Testing

System testing can be broken down into two major components - the preparation of the test cases with expected results and the execution of the tests (and correcting bugs). Usually fewer people are required to create the test cases than to execute the test. The reason for this is that a few people can generate many test cases, but during the execution, you need people not only to enter and interpret the results, but also to fix the bugs that have occurred. A rough rule of thumb is that the test case development takes 20-30% of the effort and the execution takes 70-80%. The first component, the building of the test cases, should always begin before development is complete so that once development is complete, testing can begin immediately. The amount of overlap should be kept to a minimum - 3 or 4 weeks for a 5-7 month development and 3-5 weeks for development phases greater than 7 months. You don't want to re-work some test cases because you are too close to the time they are being programmed to be certain of no changes to the specification (controlled by change requests). So, in interpreting the man months suggested from the square root method, you cannot apply the same number of people for the preparation and execution portion. You need to look at your project, decide on the number of people required for preparation, and then apply the above durations for those resources. The overall duration is approximately the one obtained by the method.

Acceptance Testing

Acceptance test time is to **support** the end users with their testing. It includes assisting in the creation of test cases and expected results, executing the tests and fixing the bugs. It is important to understand and have the Users understand that this is **their** test and the IS department is just assisting them.

The types of resources required will be analytical, to help with test cases and to explain the system, and programming to fix the bugs. Again, as in the System Test, there are two components that should be separated out - test case creation from execution of the tests and problem resolution. The former should begin during system testing and be complete just prior to the end of System Testing. Too much overlap with system testing may cause a rippling effect of problems in system testing changing acceptance test cases. Generally, you should not exceed 1/2 of the pure System Test timeframe.

Typically, the Acceptance Test involves fewer tests and fewer kinds of tests (stress test may only be done at System Testing and results provided to the User on request) and therefore should be shorter in overall duration than the System Test. One hopes also that fewer bugs will be found. As a rule of thumb, the duration of the acceptance test should be around 1/2 to 2/3 of the system test. Note, as the duration of the system test increases, the ratio tends to gravitate to the 1/2 criteria. For instance, if there were 6 month of system testing, 3 months of acceptance testing is more reasonable than 4.

All of the remaining phases typically occur in parallel with one or more of the phases above and you do not need to continue with this exercise for those phases if you are only trying to get an overall duration.

Two important concepts in determining the duration of a phase are:

1) There is a **minimum phase duration**. For any particular phase, there is a minimum amount of time that is required to accomplish the phase and it **cannot** be decreased by adding more resources.

2) There is an **ideal** phase duration. Using the **ideal** duration obtained by using the square root method, or experience, and then tuning it up or down to whole people, will **generally** provide the right duration and the right number of resources.

6.4   TUNE HIGH LEVEL GANTT

In practice, before adding the "other" phases, you should review the overall duration and see if you have been overly conservative, or if the overall timeframe is within management's expectations. Let's say there is some pressure to decrease this timeframe. How would we accomplish this?

Well let's just double the number of resources and we will get done in half the time.

Believe it or not, there are far too many people in management that believe this statement. It certainly is possible to add some resources to an "ideal" plan to decrease the duration. However, there is a limit to this that when exceeded will cause the duration to **increase**! I have personally observed the doubling of a project team during programming result in the project being further behind 8 months later than it was when additional resources were added! Knowing how many resources can be added without incurring too much risk, resulting in a productivity decrease, is an important skill to learn and exercise.

Let's review each of the phases and assess the viability and ramifications of adding resources to reduce the duration.

General Design

What are the ramifications of increasing resources to decrease this phase? To answer this, we must ask ourselves - How many distinct user areas are there to review, and how much of the analysis will be single threaded or have to be reviewed by two people? Let's assume there are two key Users and almost everything will have to go through them. Adding another resource to the analysis team may not have a positive effect, but will create a bottleneck with the two Users. Generally there is little to gain by trying to compress this very important phase - if the User feels rushed they will not sign off, or they will sign off without reviewing carefully, or completely. Both of these will result in problems later on.

Detailed Design

Detailed Design is another phase in which it is difficult to add more resources. Resources should only be added if there are a tremendous number of program specifications and they are independent of each other. Otherwise adding more resources here can result in a poorer design, causing the overall project time to increase. Caution is definitely advised if you try to add resources here.

Programming

The Programming phase is generally where the most substantive gains can be achieved. Although it is always preferable to have a small team, adding one or two persons to the ideal sized team when needed, can generally be done without negative implications.

System Testing

With System Testing it is also hard to add resources and make substantial decreases in the duration. Having too many people here will add no value. Add resources to this phase with caution.

Acceptance Testing

This phase is driven by the Users, and by the number of test cases needed to complete the test. Can it be done more quickly than our ideal plan? The answer is yes, but more onus will be on the Users. You will need to see what kind of commitment the User is prepared to provide here, and the User will probably ask how much time is required of them, so be prepared. If they understand the work involved and agree to add people here to speed up the phase, make sure that this is achievable – they are unlikely to have the background on the effect of adding resources to a plan. Many times the Users do not fully understand the requirement for Acceptance Testing and will try and convince you to decrease the duration without any more resources (they simply want to do less testing). It is IS's responsibility to ensure that there is an adequate amount of test time to complete the Acceptance Test and ensure that there is a comprehensive set of tests that will adequately exercise the systems. Given the typical User's commitment, I would decrease this phase duration as little as possible, and I probably wouldn't increase the number of IS people. Remember, implementing a system that has not been well tested and is full of bugs, will not be considered a success by anyone, and in some settings the ramifications can be severe - hospital systems poorly tested but nevertheless implemented, can affect patient safety.

Other Phases

Generally the other phases can be altered to fit the overall duration. An exception to this is the conversion phase. If there is an extensive conversion, the analysis, design, programming and testing of the conversion processes can take more elapsed time than the project itself even though there may be a lot of data entry that many people can do simultaneously. Decreasing the overall project but waiting on conversion does not make sense.

6.5   CREATE WORK PLAN FOR FIRST PHASE

Prior to creating the overall Gantt for the project, we want to confirm our estimates for the phase we are going to conduct first. This is done by creating a detailed task list, and estimating each task. Each task is usually associated with a deliverable, and for tracking purposes, it is beneficial to create a spreadsheet of the tasks and associated deliverables. The tasks and deliverables are obtained by reviewing the methodology and selecting the deliverables and tasks that are appropriate for your project.

By doing a high level task estimate, we can see how many days are allocated to each of the tasks, and potentially see problems. For instance, in General Design, we may note that we have numerous site visits required to conduct interviews, but our estimate made no allowance for travel time or there may be a great number of people who need to be interviewed. When confronted with such issues, there are two choices:

1) add time into the plan and schedule due to these abnormalities

2) determine if an alternative strategy is appropriate (like having the Users come to a Joint Application Design (JAD) session instead of individual site visits).

If there are no unusual circumstances, the estimates will be fine. This step however, forces you to think about these things now, not later.

Another common finding is the need for technical resources over and above the analysis estimate. There are specific technical tasks such as Data Modeling, Vertical Prototyping and general technical direction and guidance required by the analysts. If necessary, add the time to the Technical Architect (TA) time. Generally it is reasonable to add about 5% of the overall phase effort for the TA tasks, but it still makes sense to define the tasks and estimate them in detail. Remember to leave the General Design effort as is.

When constructing tasks and estimates there are a few guidelines to follow:

Tasks 5 Days or Less

If tasks are greater than 5 days, then you should ask yourself "Can I reduce this task to component parts?". If the answer is yes, then you should. It is easier as the Project Manager to understand the exact status of a system if all tasks are small. Often individuals indicate they are "almost finished", and yet they take much longer to finish. By keeping the tasks small, you can monitor the progress with more precision.

If a task which was 20 days long is decomposed into four 5 day tasks, and the status of the first three is complete it is more accurate to say that the fourth has 3 days remaining than to say the 20 day task has 3 days remaining because the person has a better idea of the one component that is left to complete. There are exceptions to this guideline, and if it makes no sense to decompose a task down then you shouldn't. For instance it may make no sense to break down a 7 day programming task any further.

Keep Tasks Greater than 1/2 Day

Trying to keep track of many very small tasks is a bother for you and your team. Generally, the finest level should be 1/2 day. If there are tasks that are smaller, combine them with other tasks for tracking purposes.

One Person Tasks

Try to assign individual tasks to only one person. Then it will be clear who has responsibility for completing the task. Exceptions to this are review meetings and Joint Application Design (JAD) sessions. If a person needs some assistance on a task, then schedule as few days as possible for this assistance. If the person needs more than a few days assistance, perhaps the task is assigned to the wrong person, or the task should be broken down into more tasks.

Decompose the Task

In trying to determine the exact amount of effort required for a task it is beneficial to decompose the task into its discrete elements, estimate these, and then if the tasks are too small, re-bundle them into one or two tasks.

For instance, you may have a task, "Interview User". How long should you plan for this? Break this task into its discrete elements.

- Prepare for Interview

- Interview User

- Document Interview

- Review Interview Documentation with User

Then estimate each of these finite tasks and sum them. You probably do not want to track these at this level, since they are mostly quite small, but by decomposing the task, you can see exactly where the effort is and how long to plan for each.

Determine Dependency of Tasks

Before mapping the tasks to resources on the Gantt chart, determine which tasks are dependent on each other. When you map the tasks and resources ensure that you have not violated these dependencies.

Map Critical Resources First

Usually there are one or two critical resources on a project. They are assigned tasks that others are either dependent on, or that others need to schedule their time in accordance with (reviews for instance). Always map out these resources first and then fill in the other resources.

Plan Tasks Logically

Even when there are no dependencies, it makes sense to plan tasks in an optimal manner. For instance if a person was designing or programming an "Add Client" transaction, and there was a "Change Client" transaction in the system, it would make sense to have this scheduled immediately after the "Add Client" transaction. In this way the person's time is leveraged because likely there will be edits and database accesses similar to those he has just done on the "Add Client" transaction.

Schedule Non-dependent Tasks in Parallel

Sometimes people look at the deliverables instead of the tasks involved in accomplishing the deliverables and do not schedule the tasks in parallel. For instance, the Business Case cannot be completed until the IS development costs are known, which is a task in developing the Project Plan for the next phase. However, this does not mean that all tasks associated with the Business Case need to wait until after all tasks in the Project Plan are complete. In fact, many of the tasks in the Business Case can start earlier than the start of the tasks for the Project Plan.

The following is an example of the detailed spreadsheet that should be created. It would be used as the basis for the project plan. Do not use it as a definitive plan, it was created selecting certain tasks only, and allocating hypothetical man days to them. Note the first column is the deliverable the tasks are focused on producing and the second column represents a task number.

66

DEL	Task		PM	SA	TA	Total IS	BS	BR1	BR2	Total Users
HP		**Develop Horizontal Prototype**								
	110	Develop Screen Prototypes		2		2				0
	120	Develop Report Prototypes		1.5		1.5				0
		Total	0	3.5	0	3.5	0	0	0	0
FS		**Create Functional Specifications**								
	205	Develop add client spec		1.5		1.5				0
	210	Develop add insurance spec		1		1				0
	215	Develop add mutuals spec		1.5		1.5				0
	220	Develop calculate annuity spec		2		2				0
	225	Develop change address spec		1		1				0
	230	Develop Chg deposit spec		0.5		0.5				0
	235	Develop menu spec		0.5		0.5				0
	240	Develop client by area spec		0.5		0.5				0
	245	Develop revenue by client spec		1		1				0
	250	Develop Revenue by type spec		0.5		0.5				0
	255	Develop commission by agent		1		1				0
	260	Document overall Func Spec		2		2				0
		Total	0	13	0	13	0	0	0	0
LDM		**Develop Logical Data Model**								
	305	Create Data Model		0.5		0.5				0
	310	Review and Revise		0.5		0.5				0
		Total	0	1	0	1	0	0	0	0
DDM		**Dvelop Divisional Data Model**								
	315	Revise Divisional DM		0.5		0.5				0
IS		**Create Implementation Strategies**								
	410	Develop Impact on exisit sys strategy		0.5		0.5				0
	420	Develop Training Strategy	1			1				0
	430	Develop Testing Strategy	1			1				0
	440	Review Strategies	0.5	0.5		1	0.5	0.5	0.5	1.5
	450	Finalize strategies		0.5		0.5				0
		Total	2.5	1.5	0	4	0.5	0.5	0.5	1.5
SAC		**Develop System Acceptance Critiria**								
	510	Develop SA Critiria	1			1			2	2
	520	Review Critiria	0.5			0.5	0.5		0.5	1
		Total	1.5	0	0	1.5	0.5	0	2.5	3
TR		**Document Technical Requirements**								
	610	Determine Requirements			1.5	1.5				0
	620	Document Requirements			0.5	0.5				0
		Total	0	0	2	2	0	0	0	0
BC		**Create Business Case**								
	710	Develop IS Costs	1			1				0
	720	Determine Benefits				0		2	1	3
	730	Document Business Case				0		3		3
	740	Review Buisness Case	1			1	1	1	1	3
		Total	2	0	0	2	1	6	2	9
PP		**Create Project Plan**								
	810	Develop Estimates	1	0.5		1.5				0
	820	Develop schedule	1			1				0
	830	Document Plan	2			2				0
		Total	4	0.5	0	4.5	0	0	0	0
PM		**Project Managment**								
	900	General Project Management	3			3				0
	910	Final Review of GD documents	1	1		2	1	1	1	3
	920	Revise GD documents		1		1		1		1
		Total	4	2	0	6	1	2	1	4
		GRAND TOTAL	14	22	2	38	3	8.5	6	17.5

Spreadsheet 6.3

From this spreadsheet, a Gantt chart is created for each of the resources, to ensure that the all of the tasks are completed when they need to be (the critical path), and to confirm this phase's duration. When actually mapping out each resource, often because of dependencies, tasks need to be juggled between resources

.   Since spreadsheet 6.3 is rather large and has many tasks, I'll demonstrate mapping using a smaller task plan, this one for an architecture study. The tasks and resources are depicted in spreadsheet 6.4, below

		IS resources					Total		USER		Total
		PM	TA	TS1	TS2	AA	IS		UC	UA	Users
300	DEVELOP ARCHITECTURE STRATEGY										
310	Maintain Linkage to Corporate Goals, Drivers	1	1				2				0
315	Review Enterprise Information Architecture		2			1	3			2	2
320	Review Enterprise Applications Architecture		2				2			2	2
325	Review User Interface Requirements			2			2				0
330	Review Connectivity Requirements			2			2				0
335	Review Systems Management Reqts			2			2				0
340	Review Service Infrastructure			2			2				0
345	Review Hardware				2		2				0
350	Review Operating Systems				2		2				0
355	Review Application Development and Mtce				2		2				0
360	Review Emerging Technologies, Arch, ,Stds		2	2	2	1	7				0
365	Conduct Technical Standards Review		2				2			2	2
370	Develop and Document Architectural Principles		2				2		1	1	2
375	Develop and Document Architecture Strategy		6	1	1	1	9				0
380	Identify Technological Impacts on Organization		2				2		1	1	2
385	Present Architecture Strategy	1	1				2		1	1	2
390	Revise and Accept		1				1			1	1
395	Project Management	4					4				0
	** TOTAL **	6	21	11	9	3	50		3	10	13

Spreadsheet 6.4

The first column is the task number and I have assigned task numbers to the tasks in sequential order not necessarily in order of starting. There are a few dependencies, and there are some tasks that I would like all persons assigned to begin on the same day. These tasks include tasks 360, 365, 370, 375, 385 and 390. The best place to begin is with the person who has the most tasks on the critical path. In this case it is the TA.

 At first, I list all of his tasks in the order that I want him to complete them in, and then I add all other resources. In constructing this plan, I assumed that my two TS resources could begin at the earliest on Day 4. I attempted to keep them completely busy, and ordered their tasks in the order in which they should complete them. The result is depicted in Gantt 6.1, below. Note that this Gantt is in Excel for readability; normally I do this on a simple sheet of paper, with a pencil.

	Week 1					Week 2					Week 3					Week 4					Week 5	
	1	2	3	4	5	6	7	8	9	10	11	12	13	14	15	16	17	18	19	20	21	
PM	310				395					395					395				385	395		
TA	310	315	315	320	320	360	360	365	365	370	370	375	375	375	375	375	375	380	380	385	390	
TS1				325	325	330	330	335	335	340	340	360	360	375	375							
TS2				345	345	350	350	355	355	360	360	375										
AA		315				360						375										
UC										370								380		385		
UA		315	315	320	320			365	365	370								380		385	390	

Gantt 6.1

One point about the overall look of the Gantt before addressing problems: since I didn't know the specific start date, I simply started on Day 1. Once the start date is known and the resources are known I can finalize the dates and include any statutory holidays and vacations.

As you can see, the tasks that are shaded are a problem. The TA has begun task 360 "Review Emerging Technologies, Architecture and Standards" on day 6, but TS1 can't begin this until Day 12, and TS2 can't begin this until Day 10. Yet they should be doing these tasks at the same time. Similarly tasks 365, 370 and 375 should be done at the same time, or at least started at the same time. So what do I do? Well, in this case, it is important to finish tasks 325 to 355 before 360, however, tasks 365 and 370 are not dependent on this. Therefore, I allocate tasks 365 and 370 of the TA after task 320. This almost aligns the 360 tasks of the TA, TS1 and TS2 but not quite.

There are no more tasks on the TA's timeline that I can move back. Now I must look to move TS1's tasks earlier. However, I can't start TS1 any earlier. I have two options. I can obtain another resource to do two days' worth of tasks, or I can reassign 2 days from TS1 to TA and extend the schedule by two days. Depending on the situation, either case may be the most appropriate. In this example, I'll take the 355 task from TS2 and assign it to the TA after task 370. I'll then re-adjust TS2's starting time to start on Day 8, to have continuity.

Finally, I align all the PM, UC, and UA times with those on the TA's timeline.

The result is depicted in Gantt 6.2.

	Week 1				Week 2						Week 3				Week 4					Week 5			
	1	2	3	4	5	6	7	8	9	10	11	12	13	14	15	16	17	18	19	20	21	22	23
PM	310				395					395					395					395		385	
TA	310	315	315	320	320	365	365	370	370	355	355	360	360	375	375	375	375	375	375	380	380	385	390
TS1				325	325	330	330	335	335	340	340	360	360	375	375								
TS2								345	345	350	350	360	360	375									
AA		315										360		375									
UPM								370												380		385	
UA		315	315	320	320	365	365	370												380		385	390

Gantt 6.2

Normally, it is possible to adjust the tasks among the different resources, without the schedule duration changing. However, whenever you move tasks from one resource to another, make sure that the new resource is capable of doing the tasks in the same time as the original person.

Make sure that as you make changes on the Gantt, you mark your spreadsheet with the same changes, and update the spreadsheet. As a final check, I always add up the number of days for each resource on the Gantt and compare that with the spreadsheet. Nine times out of 10, I have forgotten to adjust one or the other correctly.

6.6 CREATE LOW LEVEL GANTT (PHASE LEVEL)

We are now ready for the final step in developing the high level plan - allocating resources to the overall Gantt and ensuring we haven't used the same resource in two places, or arbitrarily decided to get more programmers for System Testing and Acceptance Testing.

Let's take each phase and map out resource types who could do the job. For General Design, map out one line for each resource and label them A1, A2 and A3, for analyst 1, 2 and 3. For the purposes of creating this overall Gantt, we need only allocate the block of time for each resource, not the detailed tasks for each phase. Hence although we would normally have a detailed Gantt for the first phase, (from step 6.5), simply summarize it on this Gantt as General Design time.

Now, map out the remaining phases. If there is a conflict, such as needing a programmer to do system test preparation, but all are busy programming, create another programmer line.

Next add in the Manuals and Training that the User will do (to get the timeframes). To add the training time, work backwards from the end of Acceptance Testing or 1 week before the end of Acceptance Testing. To add training preparation, work backwards from the start of Training. Finally to add user Manuals, work backwards from the start of User Training preparation.

As mentioned before, the bulk of Training and Manuals is for the end user. However, there may be some effort required for production documentation and maintenance training. If so, define these tasks, what the deliverables are, and estimate the effort. Typically anyone from a Programmer Analyst (PA) level to the TA could do the production documentation, and anyone from a Senior Programmer Analyst (SPA) level to the TA could do the maintenance training.

6.7 TUNE LOW LEVEL GANTT CHART

Now, we have a perfect plan. Or do we? If you look at your Gantt chart, there will likely be some problems such as:

1) Analyst Y has a 4 month gap between design and system testing.

2) Designer Z is used only during design and only for 10 weeks.

3) Analyst D has a 4 month gap between design and system testing.

4) Programmer A has not programmed anything, but comes on for system testing and acceptance testing.

5) There is a gap between programming and acceptance testing for Programmer B.

Each of the problems needs to be reviewed and appropriate action taken. Because this was our ideal plan, using ideal resources, we can make some adjustments. We can use a designer to program, we can utilize an analyst or a programmer to assist in user manuals and user training. What is important is that we review each problem, and understand the ramifications of our decisions. For instance, although we can use analysts to design and designers to program, we may not have analysts who can design and program.

As a final step, we want to add up the resource usage two ways. The first way is to simply sum the man days by resource and phase. From the Gantt, add up all of the man days for each resource. The following spreadsheet depicts the results.

Man days on low level Gantt chart									
	PM	A1	A2	TA	PA1	PA2	PA3	PA4	TOTAL
General Design		20	20						40
Detailed Design		30	30	10					70
Programming					40	40	40	40	160
System Test			30		15	15			60
Acceptance Test			15		10	10			35
Manuals						5			5
Training			5						5
Project Management	105		5						110
Technical Architect				60					60
Conversion									0
Facilities Install									0
Total	105	50	105	70	65	70	40	40	545

Spreadsheet 6.5

As a cross check and to show how we have modified our estimates from our original estimates, add a column to the spreadsheet of our synthesized phase numbers (spreadsheet 6.2), and label it "used in plan". From the low level Gantt, add up all of the days that we have allocated by phase, and enter them. From this spreadsheet we can see our original estimates based on the programming effort, the synthesized version based on our adjustments, and finally the actual figures to be used for planning and costing the project. The variances between the numbers in each column should be explained, and these reasons added to the bottom of the spreadsheet for clarity. Spreadsheet 6.6 depicts the output of the above process.

Consolidated Man days from Gantt				
PHASE	Original Project Total	New Project Total	Used in Plan	User Required Days
General Design	35	35	40	
Detailed Design	70	70	70	
Programming	159	160	160	
System Test	54	55	60	
Acceptance Test	35	35	35	
Manuals	25	5	5	20
Training	25	5	5	20
Project Management	54	55	110	
Technical Architect	32	30	60	
Conversion	0	0	0	
Facilities Install	0	0	0	
Total	490	450	545	40

Spreadsheet 6.6

7.0

PLANNING - ESTIMATING
PROJECT DEFINITION PHASE

Estimating for the Project Definition phase is slightly different than estimating General Design and onward since there are no externals yet defined. However, many of the principles and much of the process remains similar.

There are two primary components in Project Definition that are at the heart of this phase, the definition of the current process, and the conceptual design of the new solution. These two components will typically comprise the major amount of effort for this phase.

The steps for Project Definition are very similar to those for the overall project described in Chapters 5 and 6:

1) Confirm Scope and Objectives

2) Create Spreadsheet of Tasks

3) Estimate Tasks

4) Review and Revise Estimates

5) Create Work Plan

Since it is easier to discuss the tasks and estimates at the same time, I'll combine the activities of "Create Spreadsheet of Tasks" and "Estimate Tasks" into one section.

7.1 CONFIRM SCOPE AND OBJECTIVES

The scope and objectives of the project are defined at a high level in the Project Objectives Document (POD).   Depending on its depth, and the elapsed time from when it was created, the POD may accurately state the scope and objectives without requiring any tuning.   The scope can be determined by answering two questions:

1)   How many business functions are there to be reviewed?

2)   How many people are there to be interviewed?

The answers to these two questions will allow us to construct a plan to complete the Project Definition phase.

However, it's not as easy or as straightforward as sauntering up to the Business Sponsor and asking him these two questions, writing down the answers and wandering away.

What we need is a functional decomposition of the business areas that are affected by the proposed system, and a list of individuals from the different areas that are best able to define the existing system and new requirements. Depending on his knowledge of the particular business area, the IS person could begin by constructing his view of the functional decomposition, or could meet with the Business Sponsor and representatives to construct it. In either case, it needs to be reviewed and approved by the Business Sponsor. Essentially this provides us with the scope of the business areas that we need to investigate.

Next, we need to obtain the names of individuals from those business areas who can provide us with a description of the current processes, and provide us with requirements. This list should be constructed so that it is clear who in each area is to be interviewed.

Now we have a clear well defined scope - we are looking at *XX* business areas and we are interviewing *YY* people.

7.2   CREATE SPREADSHEET OF TASKS AND ESTIMATE TASKS

Create a spreadsheet that has all of the deliverables for Project Definition on the left hand side, and underneath each of these, the tasks that are required to produce the deliverable. For the purposes of the example I have used deliverables and tasks that are common to most methodologies for this phase. On the top would be the resources that are required. For resources, at this point in time, add a Project Manager (PM), a Systems Analyst (SA), a Business Sponsor (BS), and a Business Representative (BR). We may add or decrease resources later, but this is a reasonable starting point. An example of our starting spreadsheet is shown in spreadsheet 7.1.

7.2.1   Business Objectives

There are four major tasks here:

- Interview Business Representatives,
- Document Business Objectives,
- Review,
- Revise.

Other than the review, all these tasks are done by the Business Sponsor and the business representatives. It is best to obtain the Users' input on how much time they expect to expend on these tasks. Often they will allocate a block of time to do all four. Note that the review time should be separated out since IS resources should participate in this review.

The number of man days of effort for the User depends on the status of the Project Objectives Document, the background effort spent to date, the number of interviews required, and the overall size of the endeavor. Nevertheless, there needs to be a limit for the creation of the business objectives; a range 1-3 man days is reasonable, although on very large projects it may be higher.

For the review, all IS resources and all User resources should read and understand the contents. A half a day is generally adequate, and a range of 1/4 to 1 day is reasonable.

Del	Task No	Deliverable/task	PM	SA	IS TOTAL	BS	BR	USER TOTAL
BO	100	DETERMINE BUSINESS OBJECTIVES						
	110	task 1			0			0
		TOTAL	0	0	0	0	0	0
CP	200	DOCUMENT CURRENT PROCESSES						
	210	task 1			0			0
		TOTAL	0	0	0	0	0	0
SO	300	DEFINE SYSTEM OBJECTIVES						
	310	task 1			0			0
		TOTAL	0	0	0	0	0	0
CD	400	DEVELOP CONCEPTUAL DESIGN						
	410	task 1			0			0
		TOTAL	0	0	0	0	0	0
IA	500	DETERMINE IMPL ALTERNATIVES						
	510	task 1			0			0
		TOTAL	0	0	0	0	0	0
BC	600	CREATE BUSINESS CASE						
	610	task 1			0			0
		TOTAL	0	0	0	0	0	0
PP	700	DEVELOP PROJECT PLAN						
	710	task 1			0			0
		TOTAL	0	0	0	0	0	0
PM	800	PROJECT MANAGEMENT						
	810	task 1			0			0
		TOTAL	0	0	0	0	0	0
		GRAND TOTAL	0	0	0	0	0	0

Spreadsheet 7.1

7.2.2 Current Processes

The tasks required for completing the Current Processes Report are:

Review Existing System Documentation

This task involves reviewing any systems or user documentation about the system processes that support the business areas that are under review. The estimate for this effort is dependent on the amount of documentation and whether it is up to date and accurate. Because out of date and inaccurate documentation is unusable, this task is often not done. If there is useful documentation, then it makes sense to spend time reviewing it. As a *guideline*, for every 200 pages of documentation, allocate a 1/2 to 1 day for review. Make sure you look at the documentation and judge for yourself how long it will take to review.

Conduct Interviews

For each interview there are five sub tasks to be done. I have decomposed these tasks below for clarity and understanding but I would not put this detail in the overall task spreadsheet. On the overall spreadsheet, I would have one task for *each* interview.

Prepare for Interview

Prior to attending the interview, the analyst should review the area that is being analyzed to better understand the area, and thus be able to probe for more information intelligently. He may simply review some documentation, or may review with the Business Sponsor and/or business representatives each of the areas prior to the interview. Generally, 1/2 hour or less would be allocated for this.

Conduct Interview

During the interview, the analyst should be taking notes and quickly sketching rough Workflow or Data Flow Diagrams of the processes. Some basic questions are:

- What do you do?

- What does the system do? How do you interact with it?

- What is good and bad about your job/function/system?

- How could the job be done better (quicker, less errors, etc)

- What are the sources of errors?

Interviews should be brief, between one and two hours. If this is not enough time, then another interview should be scheduled.

Document

Immediately after the interview, the analyst should take the rough notes and diagrams created in the interview and create a clear comprehensive specification of that business process, making sure to obtain sample screens and reports. This task usually takes between 3 and 4 hours. The analyst should contact the interviewee to clarify any points that are unclear as he is documenting them.

Review with Interviewee

After completing the specification, time should be allocated to go back to the interviewee and review your interpretation. This is best done by walking the interviewee through the Workflows/DFD's and will take from 1/2 to 1 hour.

Revise

Any changes that were noted as a result of the review with the interviewee need to be made to finalize the spec. Generally these will take 1/2 hour or less.

If we add up all the time for these sub tasks, the range is between 5 and 8 hours. It makes sense to allocate a full day for the interview and to make sure all the sub tasks are done.

Document Current System

All of the information from individual interviews must be assimilated into one comprehensive, complete document. All of the raw material from the interviews is combined, and if there are discrepancies or clarification needed the analyst obtains the information.

The amount of time to do this is dependent on the number of processes and interviewees. Clearly if only one person was interviewed, then all the information would be the same as the data gathered at the interview, and the documenting would be just a "packaging" of the information, which should take a day or less. As a general guideline, for every 15-25 new processes that are documented, allocate one day to construct the overall Current Process report. If it is difficult to determine how many processes there are until after the interviews, as a guideline, you should allocate between 20 and 50 percent of the time allocated for the "Conduct Interview" time.

Review

As a final step, the whole Current Processes report should be reviewed with the User. Generally I would review the Workflows/DFD's, the problems and the new requirements. Some preparation time is required to create slides and ensure copies are available for everyone. The review time is again a function of how big the scope is and therefore how many processes have been analyzed and documented. One half a day for every 25-35 processes should be allocated. Once again, if the number of processes is unknown, the estimate should be in the range of 1/2 to 2 days per reviewer.

Revise

Revisions to the document as a result of the review need to be done. If a good job of analysis was done, there should be few changes. One half day or less is sufficient.

7.2.3   System Objectives

This deliverable requires some review time, some "brainstorming" and finally documenting the System Objectives. The amount of time depends on the specifics of your project, however, the task should be bound. On average it may take 1/2 to 1 day for these tasks, with an upper limit of two days.

7.2.4   Conceptual Design

There are four major tasks in developing the conceptual design:

- Conceptualize Solution,
- Document Conceptual Design,
- Review,
- Revise.

PLANNING - ESTIMATING PROJECT DEFINITION PHASE

Conceptualize Solution

To conceptualize a solution, you must take into account the current process, the strengths and weakness of the existing processes, the new requirements, the objectives and scope and then "re-engineer" the processes into a new conceptual solution.

This is best accomplished by one person if the system is small. The person(s) creates a new work flow and a new man machine boundary using workflow diagrams or DFDs. He consults with technical people, the users, and application experts as required. Often this can be a group meeting to "blue sky" different solutions and approaches.

On larger systems, a working group consisting of the Application Architect, the Technical Architect, and key Users is appropriate. This group may meet for some high level directions and then disperse while the architects architect a solution(s) and then reconvene. This may occur many times until a solution is agreed upon. In groups such as this it is very important to have a strong facilitator to ensure the group keeps moving towards the goal of defining the new solution in a timely fashion.

The estimate to accomplish this is dependent on the number of processes that need to be considered. A guideline is that for every 30-40 processes it takes 1/2 to 2 days to conceptualize the solution. This is per person, but again I emphasize that it is best to have one person, or two at the most in charge of this task, and have them bring in others when required to discuss different approaches. Group designs are lengthy and ill advised. In general I'd add 1/4 to 1/2 a day for a Technical Architect, Application Architect and a business representative. After all you want to create a solution that everyone agrees with, that meets the business goals and that is in line with the technical strategy of the corporation.

As an alternative guideline, if the number of processes is not known, the estimate for this task should be in the range of 10 to 25 percent of the time allocated for conducting interviews and documenting the Current System.

Document Conceptual Design

Once the vision is clear, it is time to document the solution. Workflow or DFD's and one paragraph descriptions of all processes are required. It is important that the man-machine boundary is clear, and that *all* externals are defined. It is from this definition that we will estimate the next phase.

As a guideline, it should take between 1/2 and 2 hours per process to document the processes and create the overall Conceptual Design document. An alternative guideline is for this task to be 30 to 80 percent of the estimate for the conduct interview and document Current System estimates.

Review

As a final step, the Conceptual Design needs to be reviewed with the Business Sponsor and business representatives to ensure that everyone agrees that this solution meets the business needs. Generally I would review the Workflows / DFD's and have the details handy if further explanation of the processes is required. There is some preparation time required to create slides and ensure copies are available for everyone. The review time is again a function of how big the scope is and therefore how many processes have been analyzed and documented. One half day for every 25-35 processes should be allocated. An alternative guideline is for this estimate to be between 1/2 and 2 days per reviewer.

Revise

Revisions to the document as a result of the review need to be done. If a good job of analysis was done, there should be few changes. One half day or less is sufficient.

7.2.5 Implementation Alternatives

Three different alternatives are to be explored at this time, but all three of them are not always required.

Build vs Buy Alternative

There are a number of tasks that could be done to accomplish this. It all depends on the degree to which you want to investigate the option, the number of packages that are to be reviewed and the amount of depth in each review. It may be a few days or a few months, if a Request for Proposals (RFP) needs to be issued and reviewed.

In constructing the estimate for this, you should delineate each task into as much detail as possible. Each alternative should have the same tasks - such as review documentation, conduct vendor presentations, etc. Then, estimate each task. Make sure it is clear how many alternatives you are reviewing and the type of review that is taking place, so that the User does not expect a full blown RFP process when you have allocated only enough time to read the "glossies".

Phases and Scope Alternatives

To produce this deliverable, there are a few interactive steps. The systems analyst must meet with the business users to determine the desired implementation dates and which functions are required upon the first implementation. Then he must schedule the tasks required and determine if it can be accomplished. If not, another meeting is required to pare down the scope, or accept the implementation date that was determined. The estimate to accomplish this depends on the number of processes that are to be developed, and hence requires scheduling. The amount of effort can be less than one day to weeks of effort. However, every attempt should be made to resolve the issues and develop the plan as quickly as possible.

Since this estimate is given before the conceptual design is done, it is most difficult to predict how much time should be allocated. If it is clear that there are business reasons to implement the system on a certain date, and if it is at all apparent that this may be difficult, 1-3 days should be planned for at this time.

Technical Alternatives

There are no real guidelines available for completing this deliverable. It really depends on what new technology alternatives are available for the specific project you are working on. At organizations that have created an Enterprise Wide Architecture, there may be few technical alternatives that need investigation. The intent should be to allocate sufficient time to investigate these alternatives when their introduction would seem to provide benefits.

7.2.6   Business Case

The creation of the Business Case is the responsibility of the Users. There are a few areas where IS is responsible. Specifically, the provision of the estimated development and operational costs is the IS's responsibility. The development costs should be developed as part of the Project Plan, and the effort to incorporate them into the Business Case should be negligible. For operational costs, the estimate depends on such factors as the number of transactions, the similarity to existing transactions, and the similarity to existing communications systems. For very large systems, an approach to take is to look at the transactions that consist of 80% to 90% of the volume of transactions, estimate these, and extrapolate the others. Typically, less than 20% of the transactions in a system make up 80-90% of the volume. This task is likely to be 1/2 to 1 day in most systems, and should be bound, with no more than 3 days allocated unless there are some unusual circumstances.

The other task that IS should be allocated to is the overall review of the business case. Typically this takes 1/2 day.

7.2.7   Project Plan

The development of a Project Plan can be broken into 4 major tasks:

- Develop Schedule,
- Document Plan,
- Review,
- Revise.

Develop Schedule

This task includes creating the task list, estimating the tasks, resolving the estimate differences, and scheduling the tasks. It is the heart of the project plan. Time should be allocated for 2 or 3 people to conduct an estimate, and for one person (typically the Project Manager), to resolve the estimates and schedule the tasks. For estimating, 1 day per person should be allocated for every 40-60 externals that need estimating. For resolving the estimate and creating the schedule, between 1/2 and 2 days for every 40-60 externals should be allocated.

Document Plan

Once the resource estimates and schedule are completed, the documentation of the plan is straightforward. By using previous plans, it should take only 1/2 to 2 days to complete this.

Review

A half day review should be scheduled with the key IS and User personnel to review the plan. This may in fact consist of two reviews, one internal with only IS, and the other with both IS and the Users.

Revise

A 1/2 day is allocated to modify the plan as a result of the review process.

7.2.8 Project Management

Time needs to be allocated for Project Management. During Project Definition, there are usually very few IS resources to manage. Generally, the Project Manager should be allocated 1/4 to 1/2 time throughout the phase's duration when there are 1-3 other IS resources on the project. If there are more, additional time needs to be allocated. If there are a number of users, or if it is anticipated that there will be more effort in managing the user resources, additional time needs to be added.

7.2.9 Sample Resource Summary

Planning required you to think about each task and deliverable for your specific project. Spreadsheet 7.2 on the following page is an example of a resource summary for a hypothetical project, with certain assumptions. Do not use this as a defacto standard for Project Definition phases.

Del	Task No	Deliverable/task	PM	SA	TA	IS TOTAL	BS	BR	OTH	USER TOTAL
BO	100	BUSINESS OBJECTIVES								
	110	Interview Business Reps				0		2		2
	120	Document Busines Objectives				0	2	1		3
	130	Review Business Case	1	1		1	1	1		1
	140	Revise Business Case				0	1			1
		TOTAL	1	1	0	1	3	3	0	6
CP	200	CURRENT PROCESSES								
	210	Review Exisitng Doc		1		1				0
	220	Conduct Interviews - acctg1		1		1			1	1
	230	Conduct Interviews - acctg2		1		1			1	1
	240	Conduct Interviews - marketing		1		1			1	1
	250	Document Current System		1		1				0
	260	Review Current Processes	1	1		1	1	1	2	3
	270	Revise Current Processes		1		1				0
		TOTAL	1	7	0	7	1	1	3	4
SO	300	SYSTEM OBJECTIVES								
	310	Document System Objectives		1		1				0
		TOTAL	0	1	0	1	0	0	0	0
CD	400	CONCEPTUAL DESIGN								
	410	Develop Conceptual Solution		1	1	2				0
	420	Document Conceptual Sol		2		2				0
	430	Review Conceptual Solution	1	1		1	1	1	2	3
	440	Revise CS		1		1				0
		TOTAL	1	4	1	5	1	1	2	3
IA	500	IMPL ALTERNATIVES								
	510	Review Phase and Scope	1			1		1		1
	520	Document Options	1			1				0
	530	Review Phases and Scope	1			1	1	1		1
	540	Revise Phase and Scope	1			1				0
		TOTAL	3	0	0	3	1	1	0	2
BC	600	BUSINESS CASE								
	610	Estimate Benefits				0	1	1	2	3
	620	Develop IS Costs	1		1	2				0
	630	Develop Other Costs				0		1		1
	640	Document Business Case				0		2		2
	650	Review Business Case	1	1		1	1	1	2	3
	660	Revise Business Case				0		1		1
		TOTAL	2	1	1	3	2	5	3	9
PP	700	PROJECT PLAN								
	710	Develop Schedule	2	1		2				0
	720	Document Plan	1			1				0
	730	Review	1	1		1	1	1		1
	740	Revise	1			1				0
		TOTAL	4	1	0	5	1	1	0	1
PM	800	PROJECT MANAGEMENT								
	810	Project Management	3			3				0
		TOTAL	3	0	0	3	0	0	0	0
		GRAND TOTAL	12	13	1	26	7	10	8	24

Spreadsheet 7.2

7.3   REVIEW AND REVISE ESTIMATES

Once the resource estimate is complete, a review should take place with the Project Director (PD). At this time, the PD may question any estimates and the Project Manager should be able to explain the estimates. The intent is to ensure that the estimates are reasonable, that the tasks are properly assigned, that as few resources as possible are used, and that the resources are used contiguously if possible.

7.4   CREATE WORK PLAN

Once the estimates have been validated, a detailed work plan should be constructed. This plan should consist of a day-by-day plan for each resource on the resource summary, including the users. An example of this day by day plan is depicted in Section 6.5.

7.5   PROJECT DEFINITION FOR PACKAGE SOLUTION

7.5.1 Preface

The methodology for Project Definition, as described previously in this chapter, holds true when the decision on selecting a package solution or building the system cannot be readily determined. Although there is typically a desire to 'buy' vs 'build', it is usually not known whether there is a package that can satisfy organizations' requirements at the time of conducting the Project Definition.

However, occasionally there are applications which an organization strongly desires a package solution, and is willing to alter some of their business practices in order to obtain a package. A good example is a Payroll or Human Resources system. Very few organizations would build their own Payroll or Human Resources system today. The packages that are available provide most if not all of the required functionality, and most have ways to add any required functionality. The price an organization pays when buying a package is less flexibility to define 'how' functions are done. Often, the package directs an organization into a particular work flow that may or may not be similar to the current work flow.

If the Business Sponsor is willing to alter the way in which work is accomplished, then a package solution can provide tremendous benefits.

Depending on the closeness of the fit, a package is usually much cheaper to develop and quicker to implement. This is not always the case, so caution is advised. The more changes that are required, the less benefit there is in selecting the package.

7.5.2 Project Definition Methodology for Package Solutions

The presumption at this point is that there **will** be a package that the Business Sponsor will agree meets the needs of the area requesting the solution. It should be recognized that if this is not true, substantial extra additional work and re-work in the Project Definition phase will be required.

Given the above explicit caveat, then the question to be answered is what do we need to produce from this phase? Rather than arbitrarily state what should be produced, let's assume that we do not have the answer, and investigate what should be altered in the methodology for this case.

Let us consider ourselves part of the organization that desires a package solution. What should we do? First, we should determine what **our** requirements are. We expect to provide information to various vendors, and have the vendors propose solutions that meet our requirements. So there are now two questions we need to answer:

1) What does a vendor require from our organization in order to respond?

2) What does our organization need in order to select the most appropriate vendor?

What does a Vendor Require?

In order to provide a solution that meets the needs of our organization, a vendor needs to know what these requirements are, and then demonstrate how their system meets these requirements.

What does our organization need?

We need to define what the requirements are so the vendor can respond, but also so that we can determine which vendor best meets these requirements (the evaluation).

All right, let's look at these requirements more closely. What exactly do we need to produce in order to provide a vendor with a clear understanding of our requirements, and for us to judge them on how well they meet these?

A vendor must know the following:

- What are the requirements of the application
- What are the technical requirements
- What are the interfaces that are required
- What is the format of the response
- What is the evaluation criteria
- What are the overall submission rules

Given that this is what they need, how can we produce it and what is the format of each area? Let's construct a more detailed Table of Contents for a Request for Proposal (RFP).

RFP for XXX Application

    1.0   Executive Summary

          1.1   Introduction

          1.2   Background

          1.3   Objectives of New System

    2.0   Instructions to Vendors

          2.1   Definitions

          2.2   Schedule of Events

          2.3   Confidentiality Statements

          2.4   Vendor's Conference

          2.5   Inquiries and Questions

2.6   Response Delivery Procedures

- when, where, how many copies

2.7   Addenda

2.8   Our organization's Right to Reject

2.9   Withdrawal of Proposal Prior to Closing Date

2.10   Disqualification of Proposals

2.11   Cost of Proposals

2.12   Required Proposal Content and Format

- includes all information that we need, such as warranties, Project Team, etc

2.13   Evaluation Criteria

2.14   Oral Presentation

3.0   User Requirements

3.1   Functional Overview

- Textual description of the overall application requirements (for vendor to get high level perspective, they will not respond to information documented here)

3.2   Functional Decomposition and Requirements

- Functional areas involved, and the requirements by area
- Requirements are stated as explicitly as possible
- State the main requirements of an area first, then add exceptions
- State volumes by function

3.3   Current User Organization

- Organization chart
- Names or at least number of key users in each functional area

4.0   Technical Requirements

3.1   Our organization's Enterprise Wide Architecture (EWA)[5] explained

4.2   Preferred Technical Environment

- Operating System
- Data Base
- Language
- Network Interface

4.3   Technical Requirements

4.3.1   Security

4.3.2   Backup

4.3.3   Recovery

4.4.4   Performance

4.4.5   Network Management

5.0   System Interfaces

5.1   Interface 1

- what data is needed
- what format
- what volumes

---

[5]See Chapter 17 for a discussion on EWA

- technical specifications/details

Appendices:

  A   Data Model

  - High level
  - if data model is not available, give table descriptions (all tables and all attributes in each)

  B   Data Dictionary

  - if not available give table descriptions (will at least show lengths)

  C   Existing Technical Environment

  - diagrams and descriptions showing existing and planned environment
  - supports section 4.1

It is now that we need to go back to the normal Project Definition Methodology and review the deliverables and the tasks and see if they are applicable, and if so in the same depth. Let's review each deliverable, one by one.

Business Objectives

All the tasks to produce this deliverable are required and the deliverable contents are the same as for a non-package solution. It is an important vehicle for not only ensuring an organization is clear on the objectives, but will provide valuable input to the vendors.

Current Process

Many of the tasks required to complete a Current Process are required, and some of the documentation is required. Specifically, the tasks required to **understand** the current business in order to convey these requirements to a vendor must be done. In essence, the objective should be to define the current system in terms of requirements for a new system, and not be concerned with the flow and the physical nature of who does what and where things are stored or passed on. However, in order to be able to list requirements of the current system, one must be assured that these have been found. The techniques of interviewing and JAD sessions using DFDs have provided an excellent method for ferreting out existing requirements (and new ones). It is easier to ask a person what he does and document this flow. When documenting the flow, holes or questions will arise that allow the interviewee to probe further and obtain the complete spectrum of the tasks.

When it comes to documenting the current system, our objective changes from the objective if this were not a package solution. We now need to convey the current requirements (and any new ones) in such a fashion that it is clear what functionality is needed, without imposing constraints that may be imbedded in the current processes. Rather than document these requirements in terms of DFDs, a different approach is required.

The approach should allow vendors to fully understand the requirements, respond with a solution, and allow our organization a straightforward capability to access which vendors meets our requirements the best. Two components are worth producing. The first would be a narrative of how the business is conducted today emphasizing the functions that are performed. This would be used to provide a perspective of what the new system must accomplish. The second section would be a list of requirements that the system must satisfy. It is easiest to construct this along the current functional model. For each major business function, a section would be created. Within this section, the requirements would be succinctly listed. All anomalies to the normal case that need to be addressed by the system need to be stated. The vendor should be required to state that they can accommodate this requirement, and, for certain requirements they should be requested to state how they accomplish this.

To summarize, the following method should be employed to accomplish the above:

1)   Conduct "normal" analysis of JADs and Interviews using the current tools and methodology.

2)   Instead of creating a complete physical description of the current system, document the functions that are required along the functional model in text form. Ensure all new requirements are included.

System Objectives

This deliverable is intended to define opportunities to apply systems technology to achieve the stated Business Objectives. As such, would it provide any benefit to our organization or the vendor to produce this? One could argue that by completing this there may be some clear technology that an organization wishes to pursue that no vendor has addressed. However, this is likely to be rare, and also violates one of the architectural principles.

This deliverable is not required, as it will not benefit an organization in creating it, nor the vendor in responding to the RFP.

Conceptual Design

This is the key deliverable from Project Definition for normal custom developed solutions. However, if we provide a Conceptual Solution to the vendors, will we assist or impede their ability to provide a solution for us.

The vendors have a solution. It may or may not fit closely with what we would produce from a Conceptual Design. However, if it does not, the vendor is unlikely to make changes to meet fundamental differences. Certainly they will either provide the tools or program custom reports. They may even provide some tools to customize the application. However, the more customization there is, the more costly the system is and the less flexible it is to new upgrades.

By providing the requirements to the vendor, we should be assured (by checking their response) that the system meets their needs. How it meets the needs will be up to the vendor. Essentially, they have done a Conceptual Design and refined it many times to their current implementation.

A Conceptual Design should not be done. Instead, a high level discussion on basic philosophies of the system should be documented. Are their specific areas that we would like to see? These could be such things as envisioning more empowered "clerks" than today, or more utilization by the end-user (the employee). Technical implications may prevent some of these, but they should be thought out and documented. One method to accomplish this is to invite one or more vendors to 1/2 day session where there is a round table discussion of what other companies are doing and what the vendor sees as strategic.

Implementation Alternatives

Build vs Buy

Clearly our intent has been to purchase a package. The component of this deliverable that is concerned with making that decision is not required. However, the activities that are concerned with creating an RFP and evaluating the responses are required. For the standard Project Definition methodology, the baseline for vendors is the Conceptual Design, but since we know we are going for a package and did not create a Conceptual Design, the baseline is the requirements that were documented during the Current System Study.

Rather than confuse the methodology with caveats and exceptions, it makes more sense to create a new deliverable - the Request for Proposal (RFP) and define the deliverable and activities separately for this.

Phases and Scope Alternatives

Depending on the size of the endeavor, this deliverable may or may not still be produced. If the conversion and implementation of the package exceeds 12 months, then some phased delivery is likely required. However, rather than define what the phases are and what is contained, the requirement should be stated that implementation is expected in xx months and if this cannot be achieved, the vendor should provide details on a phased implementation.

Technical Alternatives

The technical alternatives should not be developed. As mentioned before, the organization should define their environment and suggest that all things considered equal, vendors who are aligned with the organization's Enterprise Wide Architecture strategy will be favored over those that are not.

Business Case

This deliverable must be produced. Some components of this deliverable will be included in the RFP.

Project Plan for General Design

At the conclusion of the Project Definition phase, a package will have been selected, and a plan provided by the vendor. The organization will have some tasks that are required and may not be scheduled or tracked by the vendor, such as technical consulting, delivery oversight, conversion support or conversion programming. These are costs to the project and should be planned and tracked. Depending on the size of the endeavor a single plan from this point to implementation may be most applicable, although a plan for the next phase (as defined by the vendor) may also be applicable. For instance if the implementation will take 2 months, it makes sense to have one plan for our organization that encompasses all our tasks during that timeframe. Alternatively, if conversion of data is our responsibility, there may be 9 months of effort on this alone, which may be planned as a separate sub-project, in addition to a plan that supports other components of the vendor's plan.

8.0

PLANNING - DEVELOP
DETAILED PROJECT PLANS

8.1   OVERVIEW

Producing a realistic plan is essential for any sized project. It is the detailed plan that states who will do what, when they will do it, how long it should take, and what will be delivered. It states the organization of the project, and how the project will be managed.

Chapters 5, 6 and 7 of this document indicated how to estimate the overall project and develop an accurate high level plan for the entire project. This chapter takes us down one more level, to create a detailed plan for a phase of a project.

A detailed project plan is **the** key document for managing a project. There are many sayings that illustrate the importance of plans, such as:

> "Without a solid plan, a project is like a raft in the ocean, drifting aimlessly at the whim of the currents and wind."

Or

> "You must know where you are going, in order to recognize that you are there."

You do not want to be drifting aimlessly; you need to know where you are and where you need to get to. You need a plan.

As important as it is to have a solid plan, many projects do not have one. They may have some of the key components, such as a list of the resources, a task list by person and a schedule. However a plan needs all the components to be complete. For example, without the assumptions section, what happens if the basis for an estimate does not happen as expected? What recourse does the Project Manager have if he has not stated his assumptions for all to see, agree upon and sign off? None!

Normally a Project Plan is created for each phase. On small projects it is acceptable to create fewer plans. For a project that is 3 months duration from the start of Detailed Design to Implementation, one plan that covers all tasks across all phases would make sense. Similarly, on many projects it makes sense to create one plan covering all tasks from System Test through to Implementation. The important factor here again is for the Project Manager to think about his project's size and duration and with the Project Director's input, decide what makes sense.

In the initial phases of a project there has to be one and only one phase covered. Thus for Project Objectives, Project Definition and General Design there must be one plan each. At the end of each of these phases, a "go / no go" decision is made. The cost of that phase must be clear and only those tasks that are pertinent to that phase can be in the plan.

A plan such as the one documented in this chapter need not take a great amount of time to construct. It is important that you do take the time to complete each section of this plan, in the depth indicated. As you will find out, a number of the sections become a "cut and paste" from previous plans.

8.2   TABLE OF CONTENTS OF PLAN

The Table of Contents (TOC) defines 11 separate sections that should be completed. The following is a sample table of contents. Section 8.3 describes the purpose, contents and how to complete each of these sections.

Sample Table of Contents

1.0   Executive Summary

2.0   Deliverables Descriptions

3.0   Task Descriptions

4.0   Schedule and Milestones

5.0   Project Organization

6.0   Project Management Approach

7.0   Assumptions

8.0   Resource Estimates

9.0   Client/User Responsibilities

10.0   Financial Considerations

11.0   Detailed Work Plan

8.3   COMPLETING EACH SECTION

8.3.1 Executive Summary

Purpose

The purpose of the executive summary is to provide a summary of the plan for quick assimilation by peers and superiors.

Contents

This section includes the scope of the project, the objectives, the timeframe for this phase and the overall timeframe for completing the entire project. It should also include a financial summary of the project to date and to completion.

The executive summary should be between 1 and 5 pages in length.

How to Complete

The executive summary should be the last section created. You simply extract salient information from the different sections into a cohesive summary.

8.3.2 Deliverable Descriptions

Purpose

The purpose of this section is to clearly define each deliverable that is to be produced, so that whoever is responsible for contributing to this deliverable knows what is expected. This ensures that everyone understands what is to be delivered in this phase.

Contents

This section should start with a statement that documents the review and approval process. It should state the timeframe for review and the method for which modifications or errors will be presented back to the team. The best method is to request the User to provide a single copy with all changes marked up, or clearly identified in some other manner. The errors or proposed modifications should be clear, generally not requiring any additional information.

For each deliverable, the purpose is stated and a sample table of contents (TOC) provided. A paragraph for each section in the TOC describes the contents and the depth. The depth can be expressed in terms of number of pages, or an anticipated range of pages, that the section is intended to be. This will ensure that everyone understands the detail that will be provided, facilitating realistic expectations.

For example, if it is merely stated that a cost benefit will be done, how does anyone know what amount of work will be done or how large the document will be? The User might expect a detailed 500 page dissertation, while the estimators might expect a high level cost benefit. By defining each chapter and the number of pages, both parties will have the same expectations. The estimate for this task can then be bound, knowing that both parties understand the depth.

Each deliverable is assigned a deliverable number to facilitate cross referencing and reporting. If the organization and a sub-contractor have separate deliverables, it is easiest to refer to organization's deliverables using the first letter of the name of the organization, such as G1, G2, G3, etc, and to each sub-contractor's deliverable as the first character of their company name (or a meaningful character if there is a conflict) followed by a number.

How to Complete

Your methodology guide should provide the purpose of each deliverable that you can extract and use here. If another project had to produce this deliverable it is easiest to use that TOC, and modify it if necessary for your project. Otherwise, from the information in the methodology guide, and from what you need to deliver, create the TOC. Quantify the number of pages for each section by visualizing the section and understanding the depth that you need to provide.

8.3.3 Task Descriptions

Purpose

Each person has assigned tasks and man days associated with each task. This section is intended to provide a description of each task in sufficient detail to allow each person to conduct his assigned tasks.

Contents

For each task that is in the plan, there is a brief narrative description of what the task entails. Each task is assigned a specific task number that will facilitate cross referencing in other parts of this document.

How to Complete

Many of the tasks, such as "conduct interview" are obvious. These will be easy to document. Others require you to think exactly what you are expecting the person to do.

An example is the task defined as:

"Review Existing Documentation".

An explanation of this task may be:

"Review the Current System Specification, the Logical Data Model and the Technical Requirements document".

The goal is to be as succinct as possible but provide the necessary information for each person to understand what is expected of him in the task. You should be able to read the description and know exactly what you need to do to accomplish the task without further consultation. By being specific, the estimates will also be more accurate. Furthermore, a project review of the tasks and estimates can be conducted to validate that there is appropriate time allocated for each task.

8.3.4 Schedule and Milestones

Purpose

The purpose of this section is to provide a high level indication of the overall schedule and when milestones are planned for completion.

Contents

The schedule is a high level Gantt chart showing one bar for each phase of the overall project. If the project is in an early phase and the overall plan is not yet known, then the Gantt shows only this phase. The Milestones are listed with completion dates in a table format. See Section 6.2 of this document for an example format.

How to Complete

The low level Gantt chart is constructed when you were doing the high level plan. For the particular phase that you are planning, it may have changed slightly due to such things as exact dates for statutory holidays, etc. For this section, you only want to show the start and end dates of the phase on the Gantt, and not all of the resources. For the milestones, refer to your detailed plan and extract the dates when the milestones will be achieved.

8.3.5 Project Organization

Purpose

The purpose of this section is to clearly identify **key** personnel on the project and their relationship to the Steering Committee. This section clearly indicates the "chain of command" from both the User's side and IS's side.

Contents

A hierarchy chart is constructed showing the reporting structure and relationships among the following:

- Business Sponsor
  - o typically a senior business representative – usually a Director or Vice President level.

- User Coordinator
  - o Optional – but usually this person is assigned by the Business Sponsor to handle the day to day activities that require user management.

- IS Project Manager
  - o the day to day manager of the project from IS.

- IS Project Director
  - o usually a senior representative from the IS department with multiple project responsibilities. Typically a Director role, but sometimes is filled by the CIO on very large projects).

- Steering Committee
  - o Comprised of all of the above resources as well as senior representatives of any business area that

will be affected by the new system, and IS personnel with management responsibilities that are key to the delivery of this project.

In addition to the roles, the names of all parties are labeled on the chart.

How to Complete

There are two standard organization charts that are applicable to all projects. See Figures 8.1 and 8.2. Figure 8.1 depicts the project organization where the Business Sponsor is actively involved in day to day project activities such as handling decision requests and change requests, dealing with scope issues and other user related activities. Often a Business Sponsor has other responsibilities that prevent him from dedicating sufficient time on the project to handle these duties. In this case, he designates a person to be the User Coordinator to handle the tasks for him. This organization is shown in Figure 8.2. In either case, it is important to note that the Project Manager has responsibility for the user resources, as they relate to the project. In the case where there is a User Coordinator, he also reports to the Project Manager.

On the organization chart, the names for the key positions are labeled. It is not necessary to name the project and user team members as often they are not known at this time. The members of the Steering Committee are determined by having the key personnel meet and decide the appropriate members.

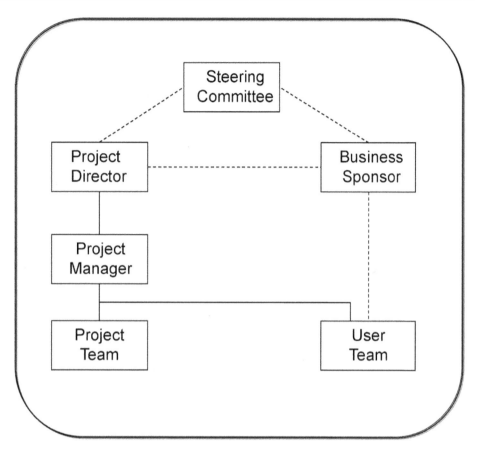

Figure 8.1

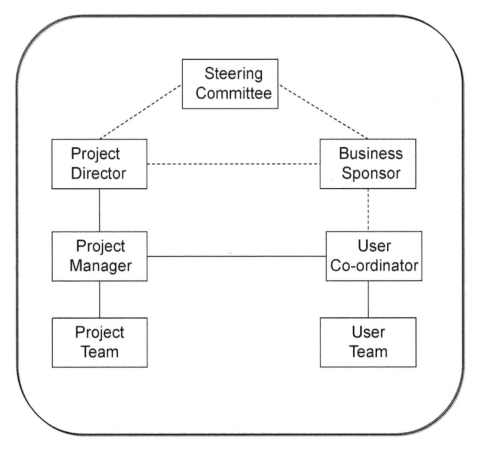

Figure 8.2

8.3.6 Project Management Approach

Purpose

To document how the project will be managed.

Contents

This section includes a discussion on:

- Project reporting (weekly, bi-weekly, Steering Committee),
- Decision request process,

- Change request process, and

- Issue resolution process.

How to Complete

Information on each of these subjects is contained in Chapter 13 of this document. It is easiest to copy the appropriate sections and delete the unnecessary verbiage. Once you have completed this section for one Project Plan, you can continually reuse it for future plans. The only component that might change from project to project is the frequency of the meetings due to the size of a project.

8.3.7 Assumptions

Purpose

To ensure common understanding and acceptance of all the assumptions that were made during the creation of the plan. The User should understand that if the assumptions become invalid, there are grounds to raise a change request which could alter the cost and schedule.

Contents

This section contains any and all of the assumptions that were made when crafting the plan. Do not include assumptions that are obvious or those that would repeat what is stated elsewhere, such as,

"John Doe will be the Project Manager".

There are actually two problems with this assumption. First, John Doe is indicated in the Project Organization section as being the Project Manager, making this statement unnecessary. Secondly, the implication is that if John Doe is not the Project Manager, then a change request can be raised to alter the schedule and/or cost. The project should not depend on any one individual.

Make sure the assumptions are valid and realistic for things that you do not have control over, but are important for the project to succeed. For example:

"Access to XYZ Server will be available from 7:00 a.m. to 7:00 p.m. Monday to Friday, and on weekends with 2 hour notice".

Not:

"The XYZ Server will be available 24 hours a day 7 days a week without any downtime".

Clearly the latter is not likely to occur, even though server availability is usually above 99%.

How to Complete

Since you are developing the plan you will be listing assumptions along the way. These form the basis for the assumptions section. There are additional assumptions such as response time and availability that you may think of as you are writing this section. It is advantageous to review other projects' assumptions to see if they are applicable to your project.

8.3.8 Resource Estimates

Purpose

To summarize the expected use of each resource for the current phase and provide a framework for resources required in future phases.

Contents

In the current phase, each resource and each task is listed with the number of man days assigned to each resource. Either a spreadsheet with the tasks on the left and resources on the top or an automated report from a Project Management tool is required here. Each task should have the task number displayed as well as the task name for easy cross referencing to the task descriptions.

For future phases, if estimates have been done, then a summary listing the current phase and remaining phases and estimates by resource is provided. This provides an overall man day estimate for the project from this point forward.

How to Complete

When creating the task lists for the project and estimating man days for each task, a spreadsheet should have been created. When completing the detailed Gantt charts, some of the man days from the original spreadsheet may have been re-assigned. When inserting the spreadsheet into the plan, it is important to verify the spreadsheet days against the Gantt chart for each resource. If an automated tool has been used, a similar report as the spreadsheet may be available. However, many automated tools do not provide the information as succinctly as a spreadsheet so you might include the spreadsheet instead. The spreadsheet provides useful information regarding load balancing and task responsibility. For instance on a spreadsheet it is easy to see where two or more people have the same or similar amount of time assigned for the same task. This would direct a Project Director to query why this is the case.

8.3.9   Client/User Responsibilities

Purpose

The purpose of this section is to ensure that any task that is outside of the Project Manager's span of control, and is clearly the User's responsibility, is documented and understood by all parties. The User should take ownership of these tasks and plan for their completion.

Contents

For each task, there is a description and timeframe when it needs to be done. For example a new computer system is to be installed at a branch office, and the office partitions and desks will be rearranged to accommodate the new workstations and workflow. Clearly, the Project Manager is not responsible for this, but someone must be. Tasks should be documented at a high level, not the detail that is developed for the IS tasks.

Other tasks may be imbedded in the plan, since there are IS resources to support them. These include tasks such as Training, Manuals and Acceptance Testing. Each of these has a brief paragraph here stating the User's responsibilities.

This section results in a list of tasks that although outside the Project Manager's control, will still need to be monitored by him. It is recommended that these tasks be monitored at each Steering Committee meeting.

How to Complete

With the business representatives, review any task or activity that is required to successfully deliver the project.

8.3.10 Financial Considerations

Purpose

To document the total cost to develop the system based on the detailed plans. This section may be omitted if all of the details are presented in the current Business Case document.

Contents

This details all project costs including personnel, hardware, software, consulting, and machine costs. This section is partitioned into one-time costs and operational costs. Historical costs from previous phases are included in order to provide a complete analysis of the project's cost.

How to Complete

The costs are summarized by phase, for each of the above mentioned components. If there are other costs not listed above, include them as line items. A spreadsheet with phases on the left and costs on the top or vice versa is recommended.

Personnel costs are a simple multiplication of the resources for a phase times the daily rate.

Hardware and software purchases are the actual purchase price in the phase when they are purchased.

Consulting costs appear in the phase in which they occur.

Licensing fees are distributed to each month and then summarized by phase. It is not necessary to be accurate in distributing the costs between two phases that are in the same month, so long as the total licensing costs for a year are accurate.

Summary totals by phase and by cost line items should be on the spreadsheet.

8.3.11 Detailed Work Plan

Purpose

The detailed work plan provides a complete schedule for completion of all tasks in the phase. It is also utilized by each resource as a model against which to track their time.

Contents

The plan contains all tasks, and for every task, who is assigned and the start and end date. Milestones are included. Gantt charts of the overall project are produced. Individual work plans for each member of the team are produced and distributed to each team member. Wherever task names are used, the task number is also included.

How to Complete

This is the heart of the plan. In constructing this detailed plan, most of the other sections fall out.

A low level Gantt chart is produced. See Chapter 6 for the details. As is indicated, often a task list is created and estimated for the current phase. Typically it is still at a high level though. Now it is time to define each task at the lowest possible level. The goal is to have no task greater than 5 days in duration, and tasks assigned to one individual only.

Many times it seems as if the task cannot be broken down. You may have a task that says "Interview User". This same task is for 2 or 3 of the analysts. But the task needs to be decomposed into the functional area that each person is involved in. A breakdown for 3 analysts doing interviews is:

"Interview Annuity Manager"

"Interview SelectPac Manager"

"Interview SelectPac Application Clerk"

These individual tasks are then assigned. By breaking down the interviews, each task is now independent of the others and can be completed at different times. Furthermore, when scheduling the tasks, the days when the SelectPac Manager is required will be shown separately from when the Annuity Manager and Application Clerk are required.

When detailing each task, all dependencies are noted and scheduled accordingly. All vacation and statutory holidays are included in the plan.

8.4 STEPS TO COMPLETING THE PLAN

There is a logical method to follow when creating a plan. There are seven major steps:

- Define Scope
- Define Deliverables
- Prepare Detailed Task List
- Allocate Resources to Estimates
- Create Detailed Gantt Chart
- Complete Remaining Sections of Plan

It is important to understand that this process is not completely serial. In fact the three processes of preparing the detailed task list, allocating resources to estimates and creating a detailed Gantt chart are done basically simultaneously, or iteratively.

Throughout this process, whenever you assume something, write it down as an assumption.

Define Scope

The very first step is to ensure that the scope of the project or phase is clear. If you are beginning the Detailed Design and all externals are not known, you should go back and complete this as part of the General Design. Deferring work into subsequent phases, especially specification of the system, is generally disastrous.

Define Deliverables

If you have proceeded with the planning as indicated in Chapter 7, you will have documented all of the deliverables you are planning to produce. If not, you should do that now. In addition, you should complete the deliverables section to the level indicated above in the Deliverables Descriptions (section 8.3.3). Point form is fine for now, the intent is to ensure that you understand the level of effort it will take to complete the deliverables, and hence allocate the right number of man days.

Prepare Detailed Task List

Once again, you should have defined all of the tasks at a higher level previous to this. You now need to decompose the higher level task into the lowest level. This can be done in a spreadsheet or with a Project Management tool. For this discussion assume it is in a spreadsheet.

The spreadsheet would have all tasks down to their lowest level on the left side, and all resources required along the top.

Allocate Resources to Estimates

Take each of the high level estimates and allocate the time to the detailed estimates and to the appropriate resource. Each of the tasks should be no longer than 5 days in duration and no less than 1/2 day. Keep in mind you have to track all of these tasks so you do not want to break any task down to less than 1/2 day.

Each task ideally is assigned to one person. If this is not possible, then it should be clear that one person has responsibility and the other is providing guidance or consulting.

If there are any "unbounded" tasks, ensure that you have bound them.

Once you have allocated all of the time to all of the resources, you may do some leveling to ensure continuity of resources.

Create Detailed Gantt Chart

For each resource, map out all tasks on a Gantt chart, taking into account holidays, vacation, and task dependencies. Because of vacations, holidays and dependencies, some juggling of tasks between resources may be required. Be sure to update the spreadsheet when you alter the Gantt.

Complete Remaining Sections of Plan

By now, you should have the two key components of the plan: the Detailed Work Plan (section 11), and the Resource Estimates (section 8). You should have as well, working papers for the Deliverables, Descriptions, Task Descriptions and Assumptions. After this, it really doesn't matter which sections you complete first, except that the Executive Summary should be written last. If the Detailed Work Plan was done using spreadsheets and manual Gantt charts, and the project is greater than two months in duration, you would now enter the data into an automated project management tool such as Microsoft Project.

9.0
PLANNING - OTHER PRINCIPLES

If you have followed the guidelines provided in the previous chapters, you ought to have a very good plan. This chapter provides a few new principles and others that need emphasizing.

9.1   COMPLETE PREVIOUS PHASE

There seems to be a tendency to create plans for future phases when all the work in the current phase is incomplete. The justification is that the next phase needs to be started right away, and we can complete what is left from the previous phase in the next phase. Depending on the phase that you are in this can cause you an undue amount of grief.

Often the tasks from certain phases are planned to be overlapped with tasks of the next phase (for example Programming and System Testing). It is possible to begin development of some programs before all are designed. These are not the phases that I speak of.

The greatest problem lies in not completing the General Design fully, and constructing a plan for the project with some "unknowns". Remember, at the end of General Design, you have a "contract" with the User to develop and implement the system for a fixed price and a fixed schedule. If there are "unknowns" that could cause you to alter your estimate, how can you do this in the next phase? The User has every right to object. Clearly all of the "unknowns" can be resolved given time. After they are resolved, a proper estimate based on facts and not conjecture can be made. It is best to spend the time in General Design and resolve these "unknowns" in this phase than to provide a plan that still has some uncertainty. This is not to suggest that you flounder in General Design for months and months and months. It is important to understand that no matter how rigorous and complete the General Design is, there will be changes. You should expect to see changes. Changes are not bad, they occur for many good reasons. Remember though, these changes must be managed by change control.

9.2   DOCUMENT ACCEPTANCE CRITERIA

When is a phase of the project over?

When does the User take acceptance of the system?

When is the project over?

These are questions that are asked all the time. They are very important and need to be answered in all project plans so that all parties agree and understand.

When is a Phase of the Project Over?

A phase of a project is over when all of the deliverables for that phase have been completed and have been approved and signed off by the appropriate personnel. Each deliverable for each phase should be documented in your methodology guide, and most often all of these deliverables are required for the specific phase. It is important to complete not only the form that is required, but also the content. For instance, if you are in General Design, one of the deliverables is the functional specifications document. Not only do you want to complete it as specified, but, you want commitment from the Business Sponsor that it meets the needs of the business unit. Signoffs without commitment are not worth much.

I have seen instances where even when the User does not understand what is being proposed, he will "sign-off" the specification. This is a result of either pressure from superiors, or fear of looking stupid for not understanding. Worse, it is often not the User's fault, - the IS people have not explained things well, have misused the techniques that are supposed to make it easy to understand! You need to ensure that your User fully understands and agrees with your approach to solving their business problems. Then all phases and especially implementation will be much smoother. There is common understanding of the end goal.

One final important point regarding deliverables - in the assumptions section of the plan, there should be an assumption that states the review and approval process for deliverables. It should state how long the user has to review the document, and how the faults are to be conveyed. Typically this is stated in terms such as:

> "For deliverables less than 50 pages there will be 2 days allocated for review. For every 50 pages in addition, an additional 2 days will be allocated. All changes to the document will be marked in one copy and returned to the Project Manager. Where specific walkthroughs of the document are appropriate, they are noted with the specific deliverable."

In this way it is clear how long the user has to review documents. If there are concerns that the timeframe indicated is too short, they can be altered now, before the plan is finalized. If the review time impacts the schedule, this can be indicated now.

When does the User take acceptance of the System?

The User should accept the system when all of the criteria that are stated in the "Acceptance Criteria" deliverable from General Design have been successfully completed. This deliverable establishes agreed to criteria that should not change.

The main thrust of the Acceptance Criteria is the Acceptance Testing. The User has agreed to limit testing to a certain number of tests and when these have been successfully performed according to the specifications from General Design, the User should accept that the functionality has been met.

The other components of the Acceptance Criteria must also be met. These may include certain technical specifications such as throughput and performance, and may also include obligations for training and manual preparation.

If you currently do not have Acceptance Criteria as a deliverable in the General Design phase you should. It allows all parties to agree up front what the rules are for accepting the system.

When is the Project over?

The project is deemed to be complete when all tasks and deliverables are complete. For all projects, this implies that the system has been successfully **acceptance tested**, moved into production, and maintenance has begun. For internally developed projects, there is typically no "warranty period" associated with the delivery, so after the system is in production, the project is officially over. If the system was developed by a contractor, there may be an on or off-site support period that is included post go-live, as well as extended warranties or maintenance agreements. In these cases, for purposes of fixing an "end date" to a project, the project is complete after the agreed upon support period has been completed.

Some projects have a "pilot" implementation, which is all of the functionality, only with a select group of users. For these projects, the project is deemed to be over when the pilot goes live, since all of the code has to have been moved to production. There should be additional project plans produced for the full rollout of the software to all users.

Some projects have a "staged" implementation, which is a portion of the functionality with all or a portion of the users. In this case, it needs to be clear in the project plan when full functionality is to be delivered. This needs to be in absolute terms in order for the plan to have the proper resources available during the "staged" implementation, and also allocated to the tasks required to complete the full implementation.

The final deliverable on any project is the Project Review deliverable. The inclusion or exclusion of this deliverable in the project plan indicates if it is part of the project or not (and hence if it is charged to the business unit or not).

9.3   DOCUMENT ASSUMPTIONS

There are two types of assumptions that a project manager makes - Project oriented and IS oriented.

Project Oriented Assumptions

The plan is based on these assumptions and if they do hold true, the plan is subject to Change Requests. These assumptions are typically stated as an expectation of someone or some entity to deliver or complete some task, where the person or entity is outside of the Project Manager's span of control.

Assumptions should be real, to the point, specific and concise. Assumptions that are sure to be false should not be stated. Some examples of valid assumptions are:

"Wiring of the Benefit Payment Offices to handle the new workstations will be done by August 30, 20xx."

"The hardware and operating systems shall be installed by ____, by July 31, 20xx."

and

"The eligibility forms will be designed, printed and distributed to the agents by September 30, 20xx."

IS Oriented Assumptions

These assumptions made by the Project Manager regarding a project cannot be subject to Change Requests if the assumption does not hold true.

Typical of these assumptions are resource expectations. The Project Manager should craft a plan with specific resource type personnel and not **specific** person reference. The Project Manager expects to have people who have the skill set he has indicated - 3 SAs, 2 SPAs, a TA and so on. These are requirements that IS management has control of and needs to provide. If they cannot provide the resource types, and provide individuals without the requisite skills to complete the job, the Project Manager should object. IS at the least should train the individual before allocating him to the project. An eager, bright, inexperienced programmer, given training, can be very productive, and meet the requirements of the job.

Another expectation of IS might be the provision of a work area and a workstation (typical for new hires).

As you can see, these are assumptions that the User should not have to worry about. They have agreed to the schedule and the man days/cost and do not expect to worry about internal IS issues.

9.4 ENSURE ALL TASKS ARE IN THE PLAN

If you have followed your methodology guide, you have created a plan that has all of the tasks that are required to deliver the system. However there are some ancillary tasks that may or may not be charged to the project, but still must be planned and scheduled.

Orientation

This is typically relevant only on very large projects where there may be one or two days allocated for a person to review previous documentation. On smaller projects, this may be handled by a one or two hour overview and does not specifically need to be entered as a task. If the task is at least 1/2 day in duration it should be scheduled into the plan.

Training

Training time may be required for many of the phases - documentation tools for General Design and Detailed Design, specific languages and the Systems Development Environment (SDE) for the Detailed Design and Programming Phase, and Test Tools for the Testing phases.

Where specific training is required prior to individuals being capable of completing their expected tasks, the plan should either show the training as part of the plan, or the people should not be allocated to the project until the training has been completed. Generally, if the person is full time on your project, and the training is to occur during the project's span, then a training task should be allocated to the individual.

Should the training time be charged to the project? This is a decision that is based on specific need versus expected level of education/training for the position. If the need is for a specific skill set not readily available within the organization, then it is appropriate that the project should bear the cost. If it is general training such as Analysis, or JAVA programming, then it likely should not. In either case the task can be allocated and tracked as part of the project. By tracking a non-chargeable task, such as training, or vacation, the individual utilization figures for a team member will accurately reflect his 100% allocation to the project, as opposed to pockets of time where it appears the person is not on the project but available.

10.0
DELIVERY - OVERVIEW

This component of the Project Model is concerned with all aspects of managing a project once it has been approved. All of Chapters 11 through 14 contain principles that apply to all phases of any project.

Let's review the Project Manager's objective:

> "To successfully deliver a project on time, on budget and to specification"

Well, this all seems simple enough, right? It isn't. Many problems arise that cause the team to **be unable** to complete the tasks, tasks both in and outside of **their** control. The result? The project falls behind, the schedule changes, the costs go up. Examples of these problems are:

- The scope can change.

- Change Requests can be late in being returned to the team.

- Decision Requests can be late in being returned to the team.

- The hardware, software and/or communications required for development are not there at the right time (be that in analysis for prototyping or all equipment for system testing).

- The hardware, software and/or communications are not working at an acceptable up-time rate.

- Deliverables are not signed off and then signed off late with changes.

- Sub-contractors do not provide their components when required.

- The User does not provide the required resources at the required time.

- Some team members are not producing as expected.

- Some users are not producing as expected.

- And, of course, Murphy's Law ("Anything that can go wrong, will go wrong") can hit whenever and wherever.

It is the Project Manager's responsibility, by utilizing and applying proper project management principles, to avoid the above problems or mitigate them in such a way that productivity of the team is not affected. Wonderful. Yes, it is true, project management is over glorified. It is the Project Manager who has to deal with all these problems. If he is successful in doing so and delivers the system on time, on budget and to specification, he must give kudos to the team, not himself. It is a thankless job.

But, the Project Manager is **the** most important person on the team. If he is successful in addressing the above problems (and any others I may have omitted), he will position the team to be successful. If he is unsuccessful, it is unlikely that any amount of application of talent by the rest of the project team will be able to deliver a system on time and on budget and to specification.

There are three primary principles that are most important in the delivery of projects:

- Controlling Scope,
- Enabling your Team, and
- Tracking and Controlling a Project.

These three important principles are addressed in Chapters 11, 12 and 13. Chapter 14 contains additional principles.

By following the principles laid out in these four chapters, you will be able to deliver your project on time, on schedule and to budget.

11.0

DELIVERY - CONTROL SCOPE

Failure to control scope is the single greatest cause of projects over-running their time schedules and budgets. The team could deliver the project on time and on budget if they only had to concern themselves with the **estimated and agreed upon work.** It is a huge concern for Project Managers to prevent scope from increasing. Let's look at the three main causes of scope control failure and how to avoid them.

11.1   BASELINE SCOPE

Clearly there needs to be a baseline from which one can ascertain if a request is within scope or not. This base should be defined in the Project Plan and should always be supported by any assumptions upon which the plan is based on. **If there is no baseline to control scope then there can be no scope control.**

The baseline is usually the Functional Specification document. Even when there is a baseline, often this baseline is not adequate. There are ambiguities, unstated assumptions, etc that prevent the Project Manager from referencing the document and saying "Clearly that is out of scope". It is important to take the time and effort (which should be planned for) to create a proper baseline that is complete and clear. If it is not, you are just delaying a problem and the problem will get worse!

In a systems integration company, I used to say that the day the contract was signed was the day the project was in trouble. The reason? The Project Manager was not writing the contract, the salesman was. It was not his concern to ensure that the scope was well defined. Even when the assumptions upon which the estimate was based were known and available, they somehow did not get included in the contract. It is extremely important that these assumptions are included. The assumptions may bound unbounded tasks, or exclude certain tasks from being part of the scope (user training for example).

What can go wrong if you *do* have a baseline document? In a classic case, there was a baseline document, the Functional Specification that was quite good. There were only 3 or 4 externals out of over 200 that were not defined exactly and without ambiguity. They all should have been, but were not. Nevertheless, it was a baseline document that stated all of the user's requirements in detail. Some years after this document was produced, the project was floundering and I was asked to look into the "scope" issue, as there were hundreds of thousands of dollars at issue. Well, I obtained the list of the "issues" and requested the Functional Specification in order that I might see what was stated there versus what the User wanted now. To my surprise, and horror, the Project Manager could not find a copy! Two days later, he obtained a copy form the User. However, this was a draft, not the final version, which unfortunately I did not discover until after my report. Many of the scope issues that were raised were extremely clear in the Functional Specification, but the Project Manager never referenced it! Clearly the Project Manager was not controlling the scope. For every scope question, the Project Manager should review the request in light of what is in the baseline document to ascertain if this in scope or should be Change Request.

Solution

What needs to be done, is to ensure the scope is well defined and known. Take extra time if required to make sure the scope is defined and the User really understands it. In the project plans clearly state what the baseline document (s) are that represent the scope, and upon which Change Control will be based. Once you have a solid baseline, use it whenever prospective changes are identified.

11.2   USER REQUESTS

Unfortunately, many of the changes to the scope are not controlled because the Project Manager does not know they have been made. Many many times I have seen Users request small (and not so small) changes directly with a programmer or analyst. Many of the programmers and analysts think that since the system is for the User, these are authorized requests and they should try and accommodate them, or the requests seem really easy and they want to please the User. Unfortunately, the result is the baseline changes without proper control. Simple changes may affect Training Plans, User Manuals, Help tips and techniques, and System Test Cases. Often the programmer does not think of these. Moreover, he has accepted more work than the estimate planned for. This can get totally out of hand and lead to major overruns.

Solution

Everyone on the project needs to understand that the only person authorized to change the functionality is the Project Manager. No one can accept the responsibility to make **any** changes that aren't approved by the Project Manager. If someone other than the Project Manager is approached to make a change, he needs to politely tell the User that "I'd like to make the change but I am not authorized, please ask the Project Manager". In this way **all** changes are controlled by the person who is responsible for the overall project.

11.3   INTERNAL CHANGES

Another common cause of increased scope is the project team itself. Sometimes people look at the solution that is proposed and believe that there is a better way to accomplish it. Rather than discuss the merits of their approach, they just go ahead and change from the specification to their solution. This problem has the same ramifications as the one above - the change may affect many different areas, and in addition, may not be the solution that the User wants and has agreed to.

Solution

It is important for the Project Manager to discuss with the team the importance of delivering the project to the agreed upon specification. In essence it is a contract with the User. It should not be changed unless there is a formal agreement to change from this specification. Therefore, if they believe strongly that the system should be different, they need to funnel this change idea to the Project Manager who can discuss this change with the User and potentially the change will be approved. Making the change without this approval violates the contract and jeopardizes the whole project success.

A procedure for accepting changes and determining whether they should be implemented must be part of the Project Manager's toolkit and must be used. Change Control does not mean that no changes are allowed. It means that the changes must be justified and approved by management before they are accepted as valid changes. Change Control is addressed in detail in section 13.5

DELIVERY - ENABLE TEAM

Enable team? Enable them to do what? The answer is simple. Enable them to do their job. In the midst of a project, the Project Manager must set aside his day-to-day tasks and concentrate first on ensuring that the team is enabled. Remove all the roadblocks before your team, and you will be on your way to success. There are two components to this:

- Ensure they can do their job,

- Ensure they are doing their job.

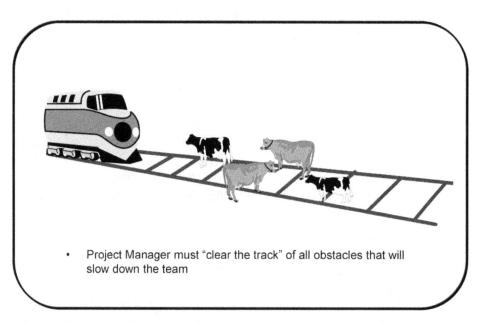

- Project Manager must "clear the track" of all obstacles that will slow down the team

Figure 12.1

12.1   ENSURE THEY CAN DO THEIR JOB

Given that the estimates for each person's tasks are reasonable and no increase of scope has occurred, the next item that causes project problems is the inability to keep the team working at 100% capacity. If you have 10 people working during programming and the internal network goes down for 1 day, preventing everyone from working, you have lost 10 man days of effort in that one day. It adds up pretty quickly, and once it is lost, there is little the Project Manager can do about it. So it is very important to ensure that it does not.

Each and every day, the Project Manager must anticipate what can go wrong today, a week from now, a month from now, 3 months from now, and make plans to either ensure it does not occur, or have backup plans once it does occur.

Too often the Project Manager doesn't hear about the problem until after it has occurred, typically at the weekly status meeting.   Joe Programmer, at the weekly status meeting, states to the Project Manager that he was unable to complete his assignment because the system was down. What should the Project Manager's response be? First he should ask more questions:

- When was it down?
- For how long?
- Why?
- Were others affected?

After this he can revise the plan if necessary, and take action to try and ensure it doesn't happen again. But could it have been prevented? Well that depends on the answers to the questions. But in general, the Project Manager has to take precautions to ensure that the environment in which people are working stays up and running on a regular predetermined schedule. Examples include requesting standard "up-time" commitments from the data center or requesting standard response times.

The first priority for the Project Manager has to be to ensure that all of the team members are not impeded in accomplishing their tasks. This means that before attending to his own work, the Project Manager should ensure that no one on the team is impeded, and if someone is, rectify the situation immediately. As long as people have the *ability* to accomplish their tasks, they will likely accomplish them on time. Although they need the educational background and training, the ability I refer to is the ability to accomplish their tasks without outside influences.

What are outside influences? Anything that prevents the person from accomplishing their tasks short of their own talents. This includes problems with hardware, software, communications, decisions, after-hours access, manuals, hot line numbers, etc, etc.

Nothing will slow your team down faster than if they are all dependent on something like a Server or a network and it goes down for long periods of time. Or, countless number of times I have heard - "I can't proceed any further until I get a decision on xxx". You as the Project Manager must anticipate these problems in advance and do your best to ensure they do not occur.

12.2   ENSURE THEY ARE DOING THEIR JOB

If people are paid to program, then they should be programming. If they are paid to analyze then they should be analyzing. However, often they are not doing what the Project Manager expects or the Project Manager has instructed them to do tasks that are best suited for others. This is most often the case with clerical or administrative tasks. Rather than utilize a talented administrative person for Visio diagrams, typing, presentation slides, etc, the person does the task himself sometimes on the instruction of the Project Manager.

Find out what employees are doing and ensure that people are utilizing their time effectively - delegate repetitive tasks downward (analysts should not be spending time doing intricate diagrams/drawings, or entering long spreadsheet data when it can be explained to another person). In one instance the Project Manager tasked an analyst to draw an organization chart, but the analyst was unfamiliar with the drawing package, so she enlisted the help of another person (who also didn't know the package). Well a simple four box organization chart took two analysts 4 hours each - 1 man day for what an administrative person could have done in 15 minutes! aughhhhhh!

At every team meeting I ask the question "Is there some clerical task that you are doing that you would rather have someone else do?". Sometimes there is, sometimes there isn't. But the team realizes that if they are doing clerical tasks that are taking away from their time doing the tasks that they need to do, that management might be able to solve this problem.

This is not meant to imply that all typing, photocopying, etc is strictly given to administrative personnel - only when it makes sense. For instance, I type reasonably well, but my handwriting is shall I say, "less than desirable". It took me longer to write down what I wanted to say, give it to a secretary, proof it, and return it to the secretary, than it did to just type it myself. However, when it came to fancy features and cosmetics, I always gave this to the administrative personnel who could do this faster and more professionally than I could.

Often the way I could tell that someone was in need of help was that he was not completing tasks on schedule, or was working long hours to accomplish tasks. Upon review, there were tasks that were slowing him down that could be reassigned. The sign of a "good" employee is not how long and hard he works, but **how smart he works**. Now if he works smart, and hard and long - you've got one hell of an employee. All team members should be leveraging their own time so that they are accomplishing the tasks assigned to them quickly and accurately.

13.0

DELIVERY - TRACK AND CONTROL

13.1 TRACK PROGRESS

Having created a solid plan for a project, one must be able to track progress against this plan. The objective of project tracking is to determine what work has been accomplished against the plan. The tracking process is used to control a project and each team member's commitment.

When the plan was created, there was a set of detailed tasks for each person, and a schedule showing when these tasks were to be started and completed. Each person on the team receives a copy of his specific tasks for the project. From this schedule and task list, the individual proceeds to complete the tasks according to the schedule unless problems arise. Prior to the weekly status meeting, each individual must determine what has been accomplished and what the **real** remaining effort will be. The time for each of the tasks an individual works on is an **estimate**. Different people work at different speeds, and some of the estimates will be high or low. It is always the case that not everything that was expected to be accomplished will be accomplished. Some tasks that were not expected to begin (or to be completed) will be.

What is **really** important is that each individual reflects his progress accurately. As the Project Manager, you have the authority when required, to juggle tasks among the resources, and potentially to add resources in order to meet the schedule. However, if you do not have the **real** picture, you will be unable to modify the plan. It is important that each individual believes it is a **team** effort and reflects accurately his progress in order for the Project Manager to take appropriate action. Each individual must **commit** to the plan of tasks for each week, indicating what he believes he will be able to accomplish in the following week.

The tracking process can be handled manually, or can utilize features of automated project management software. An automated tool will generate individual "turnaround" documents that indicate the tasks that the person is expected to work on in a particular week and the amount of time left, and will provide an area to list other tasks that may have been worked on that were not scheduled for that week.

13.2   UPDATING OF PLANS

After the weekly team meeting where each person has given his status, the plan is updated. Some tasks will show behind schedule and some ahead of schedule.

One of the challenges for the Project Manager is to deliver the project to the schedule. If some tasks are falling behind schedule, some Project Managers will reschedule the project to arrive at a new end date. The folly of this approach is that if during the next week people work very hard and overachieve, you wind up canceling the change in the schedule from the previous week. The correct approach is to always try to get back on schedule. This can be done in a few ways:

- The person falling behind should be challenged to get back on track. He may be able to work faster or even work overtime if necessary.

- If one person is ahead and another falling substantially behind such that he will not be able to catch up, then you could re-assign some of his tasks to the person who is ahead of schedule.

- Challenge all team members to over achieve in order to be positioned to take on more tasks, thereby helping the team achieve its goals.

- If no one is ahead, and some people are substantially behind, you may consider adding a resource temporarily to catch up.

When would you want to re-schedule and move the end date? As a last resort, if all of the above has failed to get you back on track, if you know you cannot catch up, then you need to reschedule. As soon as this is clear, you must convey the bad news to the User. Keep in mind, that by following the principles in this book, this should be a rare occurrence, not a common one. In fact, most projects should never need this action.

13.3 MEETINGS

13.3.1 Meetings in General

Ahhhhh, one of life's pleasures. Not really, but a necessary evil. I don't think anyone **likes** meetings, but there are a few ways to ensure that they are useful and productive:

- Create and Publish an Agenda,
- Keep Size Small
- Follow the Agenda,
- Defer Unknown Questions
- Review Results of Meeting
- Document the Meeting

Create and Publish an Agenda

Everyone who is invited to a meeting should be provided with an agenda in advance of the meeting. This allows everyone to know the objective and topics that are to be covered in order to prepare properly for the meeting,

Keep Size Small

The most productive meetings are those in which the key people are in the meeting. You want to keep the number of people to a minimum.

There is no right number of people for a meeting. However, most often, the Project Manager can represent the team effectively by himself, thus allowing the others to continue working. Based on the agenda, you can determine what topics are going to be covered and who you might need in addition to yourself. If it appears that others on your team "might" be needed and it is a lengthy meeting, have people on "stand-by". If another person's expertise is required, then ask them to come in at that time.

Follow the Agenda

Following the agenda is usually easy. What detracts from many meetings is people going off topic or speaking in too much detail for the rest of the group and for the purpose of the meeting. Both should be halted and taken off-line.

Defer Unknown Questions

In any meeting if you don't know the answer or you need further information, say so, and state that you will follow up. Then do so. Although this pertains mainly to technical questions that the Project Manager may not be sure on, it is true for any question the Project Manager feels uncomfortable answering without further investigation.

In terms of technical questions, far too often the Project Managers has got his team in trouble by committing to something he did not understand. For example, "Will all code be 64 bit?" .. "yes" .. when the answer should have been "32 bit and 64 bit where practical". By saying "I don't know, but I'll find the answer and get back to you by ___", you accomplish two things: the meeting personnel understand that you have others who are the "experts", and you do not want to commit to something you are not sure of, and your team will be happy that you have deferred a technical question to the people who are the closest to the problem/issue and know the answer. Committing your team to doing something without consulting them on the practicality and difficulty will alienate them quickly.

Review Results of Meeting

At the end of the meeting, you should determine if your objectives have been met. You may need to schedule an additional meeting if they have not. All "to dos" generated in the meeting should be stated so that team members accept the responsibility to do what you think they agreed to.

Document the Meeting

You have not completed your job if you haven't documented the results of the meeting. In most meetings there are issues raised, tasks assigned, or resolutions arrived at. It is important for all that these items are documented and distributed.

One of the better business training videos I have seen is "Meetings, Bloody Meetings", starring John Cleese. Many salient points are conveyed, and I highly recommend you view this video.

13.3.2 Team Meetings

Purpose

There are four main reasons for having team meetings:

- To determine what actually was accomplished in the previous week compared to what was planned to be accomplished,

- To solicit issues encountered,

- To obtain personal commitments for the following weeks' work, and

- To convey general information to everyone.

Who Attends?

All team members attend. If the project is very large, these meetings should be decomposed into smaller sub-project meetings. The maximum size should be around 15 people. However, if it is a large project, the whole project team should be assembled once a month to foster cross-team understanding and commitment.

What is the Structure?

The Project Manager should chair the meeting. He should request status updates and next weeks' commitments from each person sequentially. If the team meeting exceeds 5 team members, then it is best to request the status updates prior to the meeting, review them and then briefly discuss any items of concern/interest with the group. Then general information that may be of interest to all should be discussed. This could include resolved issues, unresolved issues, or global company or division news that may affect the project. Finally, each person should be queried as to whether they have any issues. Note, if there are issues that arise during the week that immediately affect progress, they should be brought to the attention of the Project Manager immediately.

How Long Should They Take?

Status meetings should be targeted to last an hour or less.

When Are They Held?

Since the status is for a week's time, they should be held late Friday afternoon, or early Monday morning. Early Monday morning is preferable since it allows for the possibility of people completing work later in the day on Friday, or even on the weekend. Meeting targets is important for individual and team morale. It is recommended that the meetings be as early on Monday as possible (8 or 9 am), before everyone begins working on that week's assignments.

Documentation

The Project Manager should keep all of the status reports, either the turnaround documents from an automated system, or his notes regarding the status and commitments for next week. Any issues should be transcribed onto the issue log. By keeping track of each week's status, the status that is presented to the User at the Bi-Weekly Status Meeting will be easily created.

13.3.3 Bi-Weekly Status Meeting

Purpose

There are two primary reasons for this meeting:

- To convey the status of the project to the User, and

- To review outstanding issues that are affecting the team and obtain resolution, or ensure the issue is assigned and will be taken care of.

Who Attends

The Project Manager, the User Coordinator, and designates of the User Coordinator attend the bi-weekly status meeting. This is a high level meeting, and often only the two need attend. However, if others are necessary, then both the Project Manager and User Coordinator may invite them. Typically the others invited are senior individuals who are able to take action on issues. For example, the head of operations may be asked to attend if there are issues regarding development or production availability and performance or if there is a sub-contractor responsible for a component, the representative of that organizations should be asked.

What is the Structure?

The Project Manager chairs the meeting. Prior to the meeting, the Project Manager prepares a bi-weekly status report that contains the current status and issues. The meeting should review all sections of this report. A sample of the bi-weekly status report and how to complete each section follows.

Highlights for Period

This section is intended to highlight major accomplishments for the reporting period. It may include such things as completion of deliverables, resolution of a major issue, or major accomplishments such as conversion of xx records, or the successful test of a major interface. Basically, if you, the Project Manager, are thrilled, overjoyed, or happy about an accomplishment, **highlight** it to the User. Good news (as well as bad) should be shared.

Tasks Scheduled for Completion this Period - Completed

These are all tasks that the team has indicated as completed in the last two weeks. This information is obtained from the Project Manager's recording of the status, or from the turnaround documents of an automated system. The actual work breakdown (WBS) code as well as the actual description from the detailed plan is used, thus allowing anyone with access to the plan the ability to find these tasks in the plan.

Tasks Scheduled for Completion this Period - Not Completed

This includes all tasks not completed this period and any outstanding tasks that are still uncompleted from the previous report. For each task, the WBS code, tasks description, resource(s) working on the task, the amount of time left to complete and the new agreed to target completion date is listed.

NAME OF PROJECT

DISTRIBUTION: XYZ Project team, Vendors, Key Users, Key IT personnel

HIGHLIGHTS FOR PERIOD:

TASKS SCHEDULED FOR COMPLETION THIS PERIOD - COMPLETED:

Task #	Description
1.1.3	Prototype Family History

TASKS SCHEDULED FOR COMPLETION THIS PERIOD - NOT COMPLETED:

Task #	Description	Resp	Hours left	New Comp Date
1.7.1.1	Create Overall Architecture	Ed B.	2	6/1/14

TASKS ONGOING (completion date spans the week's end date)

Task #	Description	Resp	On Schedule?
1.3.2.1	Develop Speech Capability for one form (No NLP)	Ross P.	yes

TASKS COMPLETED THIS PERIOD, NOT SCHEDULED:

Task #	Description
	- none

PROJECT VARIANCE

GOALS FOR NEXT PERIOD:

ISSUES:

ISSUES RESOLVED:

SUMMARY:

Tasks Ongoing

Tasks that begin in one period but do not complete in the period are listed here. For each resource that has each of these tasks, an indication of whether the task is on schedule or not is indicated.

Tasks Completed this Period - Not Scheduled

Often tasks arise that were not scheduled, but are still accomplished by the team. This may include such things as meetings with outside vendors, or meetings with other departments for clarification or input. It may also include some tasks that were completed ahead of schedule.

Project Variance

The Project Manger's interpretation as to whether we are ahead or behind schedule and by how much. If the Project is behind schedule, then this includes the Project Managers approach to getting back on schedule.

Goals for Next Period

This section includes any anticipated completion of a deliverable or milestone, or the completion internally of a deliverable (prior to User signoff).

Issues

This section is a complete list of all issues that the User has some control to resolve, or that are outside of your team's control. You want the User to know how they are impacting the team. An example might be that "there is bug in vendor xxx's software which needs fixing". These issues should be a subset of the outstanding issues that are documented in the Issue Log.

Issues Resolved

Hopefully, this includes all the issues that were reported in the above section in the previous report.

Summary

This section is the Project Manager's overall assessment of the project status. It must reflect two components. One is how many man days is the project ahead or behind schedule, and the second is the overall schedule ahead or behind. Often there can be some tasks that have not been accomplished during the period that can be done in the next period without causing the schedule to slip.

If the project is behind schedule, the summary section should be used to document how the Project Manager will get the project back on schedule. Some suggested ways are to add resources, overtime, schedule time to people who are ahead, schedule time to people who are not 100% on the project, etc.

In addition to reviewing the above status report, all outstanding and overdue Change Requests and Decision requests are reviewed.

How Long Should It Take?

The bi-weekly meeting should take an hour or less.

When Is It Held?

The bi-weekly meeting should take place soon after the weekly team meeting has been conducted. If the weekly team meeting is held Monday morning, then this meeting should be held Monday afternoon. In this way all of the status information is current.

On smaller projects of 6 months duration or less, or if the project is high profile and has risk, the bi-weekly meeting may be adjusted to be a weekly meeting.

Although bi-weekly status meetings are required, coordination between the Project Manager and the User Coordinator should occur on a more frequent basis.

13.3.4 *Steering Committee Meeting*

Purpose

The purpose of the Steering Committee meeting is:

- to provide general direction and guidance to the project,

- to resolve any major issues or policy decisions that may arise throughout the duration of the project,

- to communicate status,

- to ensure adequate resources are available to the project team as required,

- to monitor project progress,

- to ensure executive level commitment to the project is highly visible to the end user community,

- to review, and either approve or disapprove major changes that affect the scope, duration or cost of the project, and

- to expedite the approval of all project deliverables.

Who Attends?

The Steering Committee consists of senior representatives of the User Department and the IS Department who have the authority to make decisions that will affect the scope, duration, and cost of the project; generally, at least the Project Manager, the User Coordinator, the Business Sponsor, and the Project Director (the Project Managers' superior). Other senior representatives will be identified by this group for inclusion. It is important that senior representatives of all affected organization units are represented.

Many smaller projects may all be part of a single Steering Committee. If this is the case, it is typical for the Project Managers to provide their project statuses to the Project Director, and the Project Director to represent them at the Steering Committee.

Sometimes, on smaller projects, a Steering Committee is combined with the Bi-Weekly meeting.

When setting up your project's meetings, think about your project size, duration, need for User Management direction and risk and decide on the most appropriate meeting formats and schedules.

What is the Structure?

The Steering Committee meeting is co-chaired by the Business Sponsor and the Project Manager. Prior to each Steering Committee meeting, the Project Manager prepares a status report. The meeting follows a published agenda. Someone other than the Project Manager and the Business Sponsor is assigned to record the minutes. A sample agenda and description of the sections follows.

PROJECT NAME

STEERING COMMITTEE AGENDA

September 5, 20xx

1) Approval of Last Month's Minutes

2) Project Status

- Overall Summary of Progress
- Accomplished Last Month
- Planned but not Accomplished
- Milestones
- Goals for Next Month
- Decision Request Status
- Change Request Status
- Financial Status
- Issues

3) Other Topics

4) Schedule Next Meeting

Approval of Last Month's Minutes

The Chairperson requests from the group changes or additions required to last month's minutes. If there are any, they are to be recorded, and then the month's minutes approved.

Project Status

The Project Manager presents this section. He prepares slides for each of the sections and presents them.

Overall Summary of Progress

This section consists of the high level Gantt of the Project. It should include all the major phases, and be colored to show the progress as of the last meeting and now. It should serve as a reference point to where in the overall life cycle the project is currently. The Project Manager should indicate the overall status of the project as on schedule, ahead of schedule, or behind schedule.

Accomplished Last Month

The Project Manager presents highlights of major tasks accomplished in the last month. He should not present this at the level that is documented in the bi-weekly report, but rather summarize the major accomplishments.

Planned but not Accomplished

A high level view of the major tasks and activities that were not accomplished is presented.

Milestones

All milestones that were completed last month, that were scheduled to be completed by this month, and any that are to be completed in the next 3 months, are listed, with the planned date for completion, and any revised date. An example follows:

	Milestone	Original Date	Revised Completion Date	Date Completed
D4	Logical Data Model	6/8/14	6/10/14	6/10/14
D5	Program Specifications	6/12/14	6/20/14	
D6	System Test Plan	6/16/14		
D7	Programming Project Plan	9/20/14		

Figure 13.1

Goals for Next Month

This includes any major task or milestone that will be accomplished in the next month.

Decision Request Status

This is a brief numerical summary of the Decision Requests (DRs) that have been issued, that are outstanding, that are overdue, and the total number of DRs. All overdue DRs are listed (by number only), and the Project Manager should have these with him during the meeting. Each of the overdue DRs is discussed and actions plans created for quick resolution of these. An example of the summary format is shown below.

DECISION REQUESTS	ISSUED	O/S	OVERDUE	TOTAL
This Period	5	2	1	45
Last Period	10	4	0	40
Overdue Details:				
#31 - How to assess Valuations				

Figure 13.2

Change Request Status

This is a brief numerical summary of the Change Requests (CRs) that have been issued, that are outstanding, that are overdue and the total number of Change Requests. This format is identical to the format for the Decision Request summary, Figure 13.2.

All overdue CRs should be listed by number and topic. The Project Manager should have all overdue and outstanding CRs with him during the meeting and each should be discussed and if possible decisions made regarding their resolution. Note, many CRs will be resolved outside of this formal forum, in order to provide timely responses and direction to the team. Usually the Business Sponsor will solicit advice and information from those members of the Steering Committee who are affected by the change.

Once there are approved Change Requests, the log of Incorporated Change Requests should be included. See Chapter 13.5 'Change Control Management', for a full explanation of Change Requests and Incorporated Change Request logs.

Financial Status

This is a summary of costs to date, versus expected costs, and projections for project completion. It should be in graph format and depict the revised budget costs compared to the expected costs. An example follows:

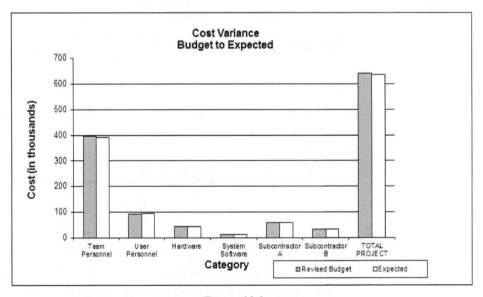

Figure 13.3

The above graph was produced from two spreadsheets appended to the written status report, but not presented. These two spreadsheets show the detailed variances by man days and by dollars and are appropriate for those individuals requiring more detailed explanations of the variances. Slides of the two are recommended to help answer any questions.

The first spreadsheet (13.1) is the man day variances:

Project Phase	Budget	Change Control	Revised Budget (RB)	Actual to date	Remain Effort	Expected Total (ET)	Variance of ET to RB
General Design	138	0	138	139	0	139	0.7%
Detail Design	227	3	230	225	0	225	-2.2%
Programming	401	9	410	132	249	381	-7.1%
System Testing	190	5	195	45	150	195	0.0%
Acceptance Test	45	0	45	0	45	45	0.0%
Manuals	10	0	10	0	10	10	0.0%
Training	10	0	10	0	10	10	0.0%
Project Mgmt	293	7	300	176	119	295	-1.7%
TOTAL	1,314	24	1,338	717	583	1,300	-2.8%

Spreadsheet 13.1

The **Budget** column shows the original estimates. The **Change Control** column includes all man days that are associated with approved Change Requests. The **Revised Budget** column includes the effect of any approved Change Requests. The **Actual to Date** column indicates the actual man days that have been expended so far. The **Remain Effort** shows the Project Manager's estimate of the remaining man days to finish the phase. The **Expected Total** is the sum of the **Actual to Date** and the **Remain Effort**. The **Variance of ET to RB** is the variance of the Revised Total to the Revised Budget. A negative number indicates that the estimates (or actuals) are under budget, while a positive number indicates the phase is over budget.

The second spreadsheet (13.2) presents the Financial Details:

Project Phase	Budget	Change Control	Revised Budget (RB)	Actual To Date	Remain Effort	Expected Total (ET)	Var of ET to RB
General Design	41,500	0	41,500	41,775	0	41,775	0.7%
Detail Design	68,200	750	68,950	62,100	0	62,100	-9.9%
Programming	120,200	2,100	122,300	39,600	84,000	123,600	1.1%
System Testing	56,950	0	56,950	13,500	41,500	55,000	-3.4%
Acceptance Test	13,500	0	13,500	0	13,500	13,500	0.0%
Manuals	3,000	0	3,000	0	3,000	3,000	0.0%
Training	3,000	0	3,000	0	3,000	3,000	0.0%
Project Mgmt	88,000	0	88,000	52,800	36,000	88,800	0.9%
Team Personnel	$394,350	$2,850	$397,200	$209,775	$181,000	390,775	-1.6%
User Personnel	$92,000		92,000	$55,000	40,000	95,000	3.3%
Hardware	$45,150		45,150	$22,332	21,500	43,832	-2.9%
Software	$12,500		12,500	$12,000	0	12,000	-4.0%
Sub-Contractor A	$60,000		60,000	$41,500	18,500	60,000	0.0%
Sub-Contractor B	$30,000	$4,500	34,500	$32,000	2,500	34,500	0.0%
TOTAL PROJECT	$634,000	$7,350	$641,350	$372,607	$263,500	636,107	-0.8%

Spreadsheet 13.2

The **Budget** column contains the original costs for the project. The **Change Control** column shows the costs that have been added due to approved Change Requests. The **Revised Budget** column is the sum of the previous two columns and represents the new, agreed to, costs of the project. The **Actual to Date** column represents the costs incurred to date. The **Remain Effort** column is the cost associated with the remaining man days for this phase. The **Expected Total** column represents the expected costs based on actuals to date and estimates to complete. The **Variance of ET to RB** column represents the variance of the Expected Total to the Revised Budget. A positive number shows an over budget category, while a negative number represents under budget.

The numbers for this spreadsheet should be generated by using the resource estimate spreadsheet's numbers, and using the automatic referencing features to extract the numbers from the resource, then applying the cost factors. In this fashion, both spreadsheets will be in sync, and there is less work for the Project

Manager, since he needs to update only one spreadsheet's resource numbers.

Issues

Any outstanding issues are presented and discussed. The issues should be presented in point form.

Other Topics

Often there are topics that a member of the committee wishes to discuss at this forum. These topics are presented by the person who asked to have them addressed at this meeting.

Schedule Next Meeting

Although the meetings should be on a set date, such as the first Thursday after the new month, the next date should be confirmed at this time.

How Long Should It Take?

Generally they should take 1-2 hours.

When Is It Held?

It is held monthly, normally on a set schedule during the first full week after month end. Since there are weekly team meetings on Monday, and bi-weekly meetings every second Monday, it is advisable to have these meetings Wednesday or Thursday in order to be able to properly prepare for them.

13.4 DECISION REQUEST MANAGEMENT

There are many decisions that must be made during the course of a typical project. Some are of a purely technical nature, such as how to design a communication module. However, decisions that affect how the business operates should be made by the User. Sometimes these decisions are arrived at via the normal analysis process. When they are not resolved via the normal process, a Decision Request should be used. Sometimes the User sees this as a tedious process. However, it is very important to utilize Decision Requests, since they document the choices available and cause thoughtful decisions to be made and documented. This is especially useful when looking back at why a system does certain things certain ways.

A Decision Request form should be used by the project team to document matters that require a decision or policy statement. It is intended to ensure that wherever a choice exists for some facet of the project, whether it is in analysis, design or procedure, the User is made aware of the decision and the decision is explained and possible recommendations made. The User is then requested to document his desire by indicating the direction in which the team is to proceed. In this manner, the User is deciding how his system will be built. These Decision Requests should be dated and responses expected back on the due date requested. These due dates should be determined in a fair manner, with consultation with the User for those that are needed in a very short timeframe. Note that Decision Request responses may result in the need for a change in scope, in which case a Change Request would be issued and authorized. The management of this process is considered key in delivering a quality system, on time and within budget.

There are two components used in Decision Request Management:

- Decision Request Log, and
- Decision Request.

The DR Log is used, as one could surmise, to log each DR. This is useful in trying to locate a DR quickly, and to ascertain the status of all DRs.

The Decision Request itself is the vehicle by which a problem or issue is documented and the resolution (decision) solicited.

The types of decisions that are requested vary in different phases.

13.4.1 Use in General Design or Project Definition

In General Design or Project Definition, many of the questions of what should occur and how it should occur are answered via the normal analysis of the process at hand. Sometimes, there is an issue that is not resolved during this analysis. These issues are usually, but not always, the result of different people's views of how something should be done. They are uncovered by the analysts in two ways:

- Through the interviewing process, two (or more) different views of how something should work arise.

- Through interviewing different persons or groups, the analyst uncovers an inconsistency.

Normally you would work with these groups to quickly resolve the issue, but sometimes the disagreement remains. Often, the analyst does not care which way to document the process, but needs to definitively know the correct, or agreed upon way that the users want the process to perform.

Both the General Design and Project Definition phases are continually uncovering decisions that need to be made on a day to day basis regarding such things as how a screen or report should look, what the logic for a process is and how should one process interact with another. Decisions such as these generally do **not** raise Decision Requests. Here are some examples of decisions that should be handled through normal analysis techniques:

- specific placement of data elements on screens

- logic for a process

- edits for fields

- sizes of data elements

Only if there is an issue that cannot be resolved through normal analysis should a Decision Request be raised.

13.4.2 Use of DRs in Subsequent Phases

The use of Decision Requests changes somewhat after the Project Definition and General Design phases are complete. At that point there should be a clear definition of the scope and an exact specification of each component of the system. So why are there any more decisions? As good a job as the User and analyst did in the previous phases, sometimes things are not as they seem. Typically either in designing a component or in programming the component, the designer or programmer determines that the component cannot be programmed as specified. In order to design and program the component, the specifications need to be altered. If the component in question is one that affects the business' agreed upon specifications, then a DR needs to be raised in order to modify that agreement.

13.4.3 Who Generates a Decision Request

DRs can be created by systems analysts, designers, programmers and the project manager. They are not to be created by the user team.

13.4.4 The Decision Request Process

The following procedure should be used when issuing Decision Requests.

1) Anyone on the team may write a DR in order to request a decision or clarification.

2) The DR is forwarded to the Project Manager.

3) The PM will review the DR to ensure that it is justified and that it is correctly completed. If it is valid and correct, the PM will assign a DR number to it, and log it in the DR log.

4) The DR will be forwarded to the User Coordinator, who will review the request, and determine which responsible person or persons can address it. If the User Coordinator determines it is not possible to achieve the target "Date Decision Required By", he will discuss the ramifications with the Project Manager immediately; otherwise he will be responsible for ensuring the DR's return on or before that date.

5) The responsible person or persons will answer the Decision Request and return it to the User Coordinator.

6) The User Coordinator will copy and file a copy of the DR for his own records, and return the original to the Project Manager.

7) The Project Manager will notify or provide a copy to the requester.

8) The Project Manager will update the DR Log with the date the DR was received.

13.4.5 Completing the Decision Request Form

Here is the step-by-step procedure for completing the Decision Request (DR), indicating the purpose of each section, what should be entered, and how it should be entered (where appropriate). Let's assume that Joe Analyst is generating this DR. A sample DR is shown on the following page.

Request Name

The purpose of this is to provide a succinct name for the Decision Request, one which indicates the topic of the DR solely by the log entry. Many times you will want to scan the log looking for a particular DR, rather than look through each DR in detail.

Number

The purpose of the Decision Request number is to provide a means to track and reference each DR.

Joe Analyst leaves this blank, since the Project Manager controls these, via his log.

Requester

The requester is the person initiating the request. It is included so that upon return the DR is copied to the person who wanted the information.

Joe enters his name here.

Date Decision Required By

This is the latest date that the requester can wait for the decision. Many projects have targeted return dates of 3 to 5 days for all Decision Requests. However, sometimes urgent DRs require quicker turnaround.

Joe determines when he needs the information and enters this date.

Project

This identifies the specific project to which the Decision Request refers. In large organizations, multiple projects could generate multiple DRs to the same user community, and this serves as identification of the system.

Decision Request	
Request Name:	**Number:**
Requestor:	**Date Decision Required by:**
Project:	**Issue Date:**
Decision Requested:	
Recommendation / Impact:	
Reply to Decision Request:	
User Authorization:	
Name:	**Date:**
Signature:	

Issue Date

This is the date the Decision Request was issued. It is used to track when the request was initially made.

Joe enters the current date.

Decision Requested

Now the important stuff. The purpose of this section is to succinctly request a decision by the User.

By the very nature of the term used to describe it, it would seem perfectly obvious that a **decision** on something should be **requested.** I make this point because as basic as it seems, many people make it very difficult for themselves and the project by not writing Decision Requests properly.

All Decision Requests should **succinctly** state the following:

- The reason a DR was raised. Was it due to a required change to a specification in order for the process to work? Was it due to two or more different views? Be succinct, but be specific - who differed? (e.g. "The Functional Specifications document on page 7, paragraph 3 states ..., however, ... cannot be done because of ...")

- The decision that is required should be briefly described. For example, "We need to determine how to calculate ..., when ... occurs"

- The options available should be numbered and described.

- The final statement should be a question, such as "please choose one of the above three options". There should be only one question posed in the DR and it should be **clear what the question is asking**.

Recommendation / Impact

One of the above options is selected, and the reasons for it supported. Any impact that may arise out of choosing this option is stated. In particular, will following the option cause a Change Request to be generated that will cost more time and/or money?

Joe enters his recommendation here, following the above guidelines.

Reply to Decision Request

The purpose here is to have the User's decision in writing.

The User Coordinator indicates his decision. Typically the answer is simply "option 2", or "Agree with recommendation". If it is difficult for the User to state his decision succinctly, the wording of the request is probably poor. Make it easy for the User to answer!

User Authorization

Formal approval with a date and a signature will ensure that the correct process has been followed and the User has authorized the decision.

The User Coordinator dates and signs this.

13.4.6   Completing the Decision Request Log

Let's review the process for entering information into the DR Log. A sample of the DR Log is on the following page.

At the outset of the project the log is created and the **Project Name** and **Project Code** are entered.

Number

This column contains the Decision Request numbers for previously entered DRs. When a new DR is created, the Project Manager enters the next sequential number on the log

Initials of Requester

The purpose of this column is to quickly find a DR generated by a certain individual. The initials of that person are entered here.

Issue Date

This is the issue date on the Decision Request.

Request Name

This is the Request Name on the Decision Request.

Decision Request Log							
Project Name:					**Project. Code:**	**Page**	**of**
No	**Init of Reqstr**	**Issue Date**	**Request Name**	**Date Required**	**Date Received**	**Change Request Number**	

Date Required

This is the "Date Decision Required By" field on the Decision Request. The Project Manager can quickly find outstanding DRs by scanning for blank dates in the "Date Received" column of the log and comparing the current date with the "Date Required".

Date Received

This is the date on which the Project Manager receives the answered Decision Request.

Change Request Number

> Should a Decision Request cause a Change Request to be generated, it is useful to have a cross reference from the Decision Request to the Change Request.

13.5   CHANGE CONTROL MANAGEMENT

It is imperative that all people involved on a project are aware of the development scope and objectives at all times. If there is a need to change any item that could impact the scope, objectives, schedule or effort required, then it must be documented and approved before being incorporated.

Change control does not mean that no changes are allowed. Rather, in recognition that changes will occur, it is a method of reviewing and understanding the impact of potential changes prior to accepting them. The control of changes to the scope of the project is an essential ingredient in the success of a project. All possible effort must be made to ensure that changes are instituted only where and when necessary, and that the impact of the change on the project schedule and project cost is understood. Often, non-critical changes are deferred until after implementation, thereby not adversely affecting the implementation date.

The primary purposes of the Change Control procedures are:

- to schedule the resources to deal with the requests;

- to monitor the progress of the task;

- to ensure that the relative importance of each request is evaluated and assigned a corresponding priority; and

- to accumulate enhancement requests that will be incorporated in a subsequent phase.

There are three components used in Change Control Management:

- Change Request Log,

- Change Request, and

- Log of Incorporated Change Requests.

13.5.1 Change Request Process

The following procedure is used in conjunction with the issuance of Change Requests.

1) A request for a Change is generally generated by someone in the user community. The User Coordinator is responsible for ensuring the change is justified and requesting that the team respond.

2) The Project Manager logs the Change Request in the Change Request Log.

3) The Project Manager and the team review the Change Request and determine the magnitude and impact of the change in terms of man days of effort, schedule changes, and cost.

4) The response to the Change Request is forwarded to the User Coordinator.

5) For "no cost" changes requested by the User and affecting neither cost nor schedule, the User Coordinator usually solicits sign off from the Business Sponsor.

6) For Change Requests that affect the cost and schedule but with no urgency to the decision, the CR is tabled as an issue to be resolved at the next Steering Committee meeting.

7) For CRs that need a quick response, the User Coordinator typically convenes a meeting of all or some of the Users who are part of the Steering Committee and decide on the outcome.

8) The signed-off Change Request is returned to the Project Manager. He updates the Change Request Log with the returned date and the resolution.

9) For "approved" Change Requests, the Project Manager updates the "Log of Incorporated Change Requests" and his Project Plan.

13.5.2 Completing the Change Request Form

Let's follow the procedure for completing the Change Request: the purpose of each section, what should be entered, and how it should be entered (where appropriate). A sample Change Request follows.

Change Request		
Request Name:		**Number:**
Project:		**Issue Date:**
Description of Change:		
Estimate of Cost:		
Impact of Change:		
Recommendation		
User Authorization:		
___ Approved ___ Withdrawn ___ Deferred	**Name:**	**Date:**
	Signature:	

Request Name

The purpose of this is to provide a succinct name for each Change Request so that the topic of the Change Request is obvious from the log entry. Many times you will want to scan the log looking for a particular Change Request, rather than looking through each Change Request.

Number

The Change Request number is used to track and refer to each Change Request.

The Project Manager assigns the number when he receives a Change Request from the User Coordinator.

Project

This identifies the specific project to which the Change Request refers. In large organizations, multiple projects generate multiple Change Requests to the same user community, and this identifies the system.

Issue Date

The issue date is used to track when the request was initially made

Description of Change

This is detailed request for a change. Assuming you are in a phase beyond General Design, it should be as detailed as the Functional Specifications. Remember, this is a change to an agreed upon specification, and you want to be as rigorous in understanding the change as you were in specifying the original scope. Indeed, on major changes, this could result in a mini-Functional Specification.

Estimate of Cost

This is detailed as well. Costs by phase, by task and resource should have been created, and need only be summarized by resource type in this section. The backup material remains attached to your copy of the Change Request.

Impact of Change

This section indicates any schedule or organization changes that will be impacted by this change.

Recommendation

The Project Manager makes a recommendation for one of the three options (Approve, Withdraw, or Defer) only when there is a particular reason to do so. Many times it doesn't matter to the team if a change is added or not. However, depending on the phase, morale may be lowered if a major Change Request causes them to back track out of System Testing into Programming. On the other hand, sometimes a Project Manager may want to have a Change Request approved because it will be easier to implement the system.

User Authorization

The purpose here is to obtain the Users decision in writing. The Business Sponsor would indicate his decision by checking the appropriate selection box and signing the Change Request.

13.5.3 Completing the Change Request Log

At the outset of the project the Project Manager creates the log on a Word Processor or Spreadsheet, and enters the **Project Name** and **Project Code.** A sample of the Change Request Log follows.

Number

This column contains the Change Request numbers for previously entered Change Requests. When a new Change Request is created, the Project Manager enters the next sequential number on the log.

Issue Date

This is the issue date on the Change Request.

Request Name

This is the Request Name on the Change Request.

Change Request Log

Project Name:			Project. Code:		Page of	

No.	Issue Date	Request Name:	Est. Cost	Status	Date Closed

Estimated Cost

This is the estimated cost in dollars that is associated with the Change Request. It is the total in the "Estimate of Cost" area of the Change Request.

Status

This is status of the Change Request. A blank status implies the Change Request has not yet been resolved. Once a Change Request is resolved, this resolution is updated to one of the three on the bottom of the Change Request.

Date Closed

This is the date on which the Project Manager receives the signed-off Change Request.

13.5.4 *Completing the Log of Incorporated CRs*

Every incorporated Change Request is entered in this log. The log provides an ongoing, running total of what the **accepted** project cost and durations currently are. An example of a log follows:

INCORPORATED CHANGE REQUESTS								
CHANGE REQUEST		ORIG	NEW	CHANGE		REVISED PROJECT		
No	Description	EST	EST	Days	$	Days	Cost	Date
	ORIGINAL					3,100	$2,995,000	5/21/2013
1	Change Annuity Trx	20	25	5	$2,500	3,105	$2,997,500	5/21/2013
2	Add Dental Calc		500	500	$250,000	3,605	$3,247,500	8/21/2013

Spreadsheet 13.3

NO

This is the approved Change Request's number.

Description

This is the Change Request's Description.

Orig Est

This column is used when there is a change to an existing function. It shows the original man day estimate for this function.

New Est

This is the new estimate for a function. If the function did not exist before, then this shows the man days that have been estimated.

Change Days

This is the total man day change as a result of the Change Request.

Change $

This is the dollar cost effect of the Change. It represents the net increase or decrease to the cost of the project.

Revised Project Days

This represents the new total number of man days for the complete project.

Revised Project Cost

This is the revised total project cost.

Revised Project Date

This is the revised **end date** of the project.

Note, the first entry should always be the Original days, cost and end date, to put the future changes into perspective.

13.6 ISSUE MANAGMENT

Many issues arise throughout the course of a project. As stressed in Chapter 12, Delivery - Enable Team, the effective resolution of issues that can constrain or even halt the progress of the team is very important. Issues arise at scheduled meetings, and during every day work. It is necessary to have a formal method to record and track the issues and their resolution. Yellow stickies with scribbled issues tend to get lost!

13.6.1 The Issue Process

The following procedure should be used when issuing Issue Forms.

1)   Anyone on the team may write an Issue Form to request direction, or to identify an area that needs addressing by someone.

2)   The Issue Form is forwarded to the Project Manager.

3)   The Project Manager will review the Issue to ensure that it is

clear and correctly completed. If it is valid the PM will assign an Issue number and log it in the Issue Log.

4) If the "Issued to" area is completed, the Project Manager will determine if he agrees with this and if so, forward the issue to that person. If it is not completed, the PM determines who will resolve the issue and forwards the Issue to that person.

5) The responsible person will respond to the Issue and return it to the Project Manager.

6) The Project Manager will review the resolution and forward a copy to those listed on the distribution list.

7) The Project Manager will update the Issue log with the date the Issue was received.

13.6.2   Completing the Issue Form

Here is the step by step procedure for completing the Issue Form, which is shown on the following page.

Issue

The purpose of this is to provide a succinct name for the Issue, one which indicates the topic of the Issue solely by the log entry. Many times you will want to scan the log looking for a particular Issue rather than look through each Issue.

Number

The purpose of the Issue number is to provide a means to track and reference each Issue. The requester should not include a number; this should be done by the Project Manager once he receives it.

Requester

The requester is the person initiating the request. It is included so that upon return the Issue resolution is conveyed to this person.

Issue Form	
Issue:	**Number:**
Requestor:	**Date Decision Required by:**
Issued to:	
Project:	**Issue Date:**
Issue Description:	
Issue Resolution:	
Issue Resolved by:	
Distribution:	**Auth Date:**
Authorization Signature:	

Date Decision Required By

This is the latest date that the requester can wait for the decision. Many projects have targeted return dates of 3 to 5 days for all Issues. However, sometimes urgent Issues require quicker turnaround. The requester should provide the latest date that is acceptable, such that the project is not delayed.

Issued to

The individual who the requester believes should review the Issue. If the individual is unknown, this could be the department.

Project

This identifies the specific project to which this Issue refers. For issues that are of an overall global nature, this should be stated as "ALL".

Issue Date

This is the date the Issue was generated. It is used to track when the request was initially made.

Issue Description

The purpose of this section is to succinctly request a resolution to an issue.

Many *Issue Descriptions* have three parts,

- The Issue that requires a resolution should be briefly described. For example, "We need to determine whether ... is in scope or not"

- If the requester knows of options, these should be numbered and described.

- The final statement should be a question, such as "please choose one of the above three options". There should be only one question poised in the Issue Request and it should be clear what the question is asking.

Each description needs as a minimum the first and third parts.

Issue Resolution

The purpose here is to respond and resolve the issue. It should be a clear concise answer to the issue, if it is difficult for the person to answer, the original issue may be poorly worded.

Issue Resolved by

The name of the individual who resolved the Issue is indicated.

Distribution

Other project teams or users may be interested in the resolution of an Issue. This list indicates who besides the requester is to receive a copy of the resolved Issue. It is completed by the requester and the Project Manager may add additional teams or individuals.

Authorization

Formal approval with a date and a signature will ensure that the correct process has been followed and the User has authorized the decision.

13.6.3 Completing the Issue Log

The form, a sample of which follows, is basically self-evident except for the. "RESP" column. This column is for the initials or name of the person responsible for resolving the issue. Often the issue needs someone other than the Project Manager to resolve it; for example, resolving a technical issue with a vendor.

Issue Log

Project Name:		Project. Code:	Page of	

No.	Description	Date Added	Resp	Date Resolved

14.0

DELIVERY - OTHER PRINCIPLES

14.1   PROJECT DELIVERABLES

Deliverable quality reflects the quality of the system being produced. Deliverables should look professional. Nothing puts off a reader more than spelling errors throughout the document. With the existence of "spell checkers" in all word processors, this should never happen. While the goal is to produce quality deliverables, one should not get trapped in writing the great American novel, nor worry about the prose. Keep sentences simple, and to the point.

Before a deliverable is released to the User for review, an internal review should always take place. This review should ensure that the deliverable has all necessary sections, and meets the expectations set out in the deliverable list in the project plan. In addition, the review should ensure that the deliverable is understandable by the target audience. This often means reviewing the deliverable by placing yourself in the role of the reviewer, and asking the question "Could I understand this if I had the knowledge of the reviewer?". Quite often, the author is so involved in the project and technology that he takes too much for granted, and as a result loses his audience.

The Project Manager should review every deliverable before it is sent out for approval. As the person with the overall responsibility for the project, he must review each deliverable to ensure that it meets the criteria set out in the Project Plan, and that it is a quality deliverable. Others who may review deliverables include the Application Architect and the Technical Architect. This is not to say that they are the only ones responsible for the quality. It is the individual who is producing the deliverable to produce a quality document that should be approved both internally by the project team and the end user.

14.2   PROJECT LIBRARY

The purpose of a project library is to provide a central site where all deliverables can be found. They need to be found not only for the project team to reference, but also for other project teams to review and use.

Many deliverables are similar from project to project. Certainly the format and some of the verbiage of a Project Plan can be used on all projects. Wherever possible I try and use an existing deliverable and modify it for my particular project specifics.

Normally I keep the originals of all deliverables on a shareable drive or in a tool such as Sharepoint, both which allow authorized users to access these documents. Make sure that wherever they are stored that this is being backed up regularly (daily).

14.3   PROJECT HANDBOOK

The Project Handbook comprises all the information the Project Manager needs to manage the project. It is a one-stop reference handbook that not only the Project Manager can use, but anyone wishing to find a Change Request or Decision Request, or requiring a form. Conceptually it is referred to as one handbook, but invariably, it consists of many physical or electronic documents. All of the components should be stored in close proximity to the Project Handbook so that all may find the information when needed.

The following is a complete list of all sections that are in the Project Handbook. Not all sections are relevant to all projects, but every project should start with this list and then delete or add as appropriate.

PROJECT HANDBOOK

1. Project Scope and Objectives

2. Contract, Agreements, Proposals

3. Project Plans

4. Resourcing

5. Finances

6. Outstanding Issues List

7. Change Control Components

8. Decision Request Log and Requests

9. Acceptance Documents

10. Project Forms

11. Project Correspondence

12. Steering Committee Minutes

13. Bi-Weekly Status Reports

14. Weekly Individual's Status Reports

15. Project Statistics

16. Timesheets

17. Computer Costs

18. Project Highlights

14.4   USER COORDINATOR ROLES AND RESPONSIBILITIES

The role of a User Coordinator can be full time or part time depending on the project's size and complexity. It may be filled by the Business Sponsor, or by his designate. The coordinator serves as an intermediary to the user community.

Liaison between IS and User community

In this capacity, the coordinator is responsible for finding answers from the User community for questions presented by the Project Manager. The question may be a formal Decision Request, a Change Request or just a verbal question. It is the User Coordinator's responsibility to find the individuals who are required to answer the question and have them answer it. In the case of a formal Decision Request, there may be many User personnel who need to provide input.   The User Coordinator ensures that the proper personnel have input and then provides a single response to the IS Project Manager. Often there is not an immediate agreement on the approach. It is the User Coordinator's responsibility to solicit the responses and obtain consensus, or if that fails, to make a decision that is supported by the Business Sponsor.

First Level Management of User Project Team

Although the Project Manager has overall control of the User's time as it relates to the project, often the first level of managing their time is controlled by the User Coordinator. This may include collecting status, reporting status, and initial problem resolution. However, this is not mandatory for this role.

Signing Authority

Typically the User Coordinator has the authority to approve Change Requests and Decision Requests. If the User Coordinator is not the Business Sponsor, then there needs to be clear direction from the Business Sponsor as to what his authorization level consists of. The IS Project Manager must understand the level of authority provided to the User Coordinator. The best way to do this is to have the level clearly documented in the Project Plan, which is approved by the Business Sponsor. However, most User Coordinators will interact with the Business Sponsor on all Changes that affect the cost or the duration of the project.

15.0

PROJECT REVIEWS / AUDITS

Reviewing different aspects of the project at different times is a healthy process. It can uncover some areas that need improvement before they become an issue, or in a post-project review, they can identify areas that need to be addressed in future projects. The terms "project audit" and "project review" are to be considered the same, some organizations prefer to not use the word "audit" as it may have a negative connotation.

Prior to beginning any phase, there should be a formal review. This review includes at a minimum the Project Manager, Project Director and Business Sponsor. What happens once the project begins? Are there no more formal reviews? In many organizations there are not, however, there should be more formal reviews on all projects.

What is the purpose of reviews in general? Sometimes projects begin to go off-track due to oversight of certain components, or the project team being too close to the situation and "not seeing the forest for the trees". Project Reviews serve as regular check-ups on the health and welfare of a project.

Are reviews witch hunts designed to seek out the people who have done poorly, to lay blame wherever possible, to misrepresent the truth? If done incorrectly or with this intention they can be, and I have seen this. When this occurs, reviews become useless, and indeed harmful to the organization. People will be reticent to tell the truth for fear of retribution. It is to the detriment of the project, to the team and to the Project Review process to make statements unsupported by facts.

The functional purpose of a review is to *improve the process.* Without review, no one will know what worked well and what didn't. Reviews provide input into the process of improvement by providing accurate measurements and concrete and subjective observations about the process. Then IS can determine what the

173

reasons for success and failure were, ensure that successful components are propagated throughout the organization, and determine solutions to eliminate the failures.

Reviews should be thought of as positive processes. The reviewer must take the time to ensure the truth is being stated, and to state it in a non-threatening manner. Reviews need to emphasize "what went well" as well as what didn't. This is not to say they should be candy-coated. The information in reviews needs to be based on facts and needs to be detailed enough that management can understand the problems and successes and take action. Nebulous statements provide no benefit.

15.1 PHASE REVIEW

Phase reviews occur at the end of a phase, or when the current project plan has been completed (since some plans may encompass more than one phase). Phase reviews should review how well the plan was executed. One of the best measures is comparison of the estimates for each task with the actual amount of time spent. Whether this is interviewing time or programming of individual modules, the comparison should take place at the most detailed level. From this, one can determine where the estimates were accurate and where there were problems. Once this has been determined, a more detailed review of the problems should be done to determine the cause. Were the estimates low? Was the individual not performing at expected levels? Was the scope accurate? Depending on the answers, different actions are taken; for example, providing more education to individuals, or altering the estimating parameters.

One might alter the estimating parameters if consistently the effort to conduct some activity takes longer than the time allocated. For this to occur, you need to have constructed estimating metrics from which you can compare the actuals to the estimated value. For instance, if an estimate to program a simple Java program was three days, but the actuals over the course of time suggest the estimate would be more accurate at four days, the estimating metric for a simple Java program should change to four days.

Another important measurement is the satisfaction of the User. Was there anything that could have been done to improve the process from the User's perspective? This requires a brief interview of the Business Sponsor, the User Coordinator, and all or some of the Business Representatives (two or so is sufficient).

A one or two page summary is documented by the Project Manager identifying the successes and failures, and recommending action for the problem areas. This may result in recommendations regarding estimates, or any facet of the methodology including Business cases.   If the Project Manager has been tracking the progress accurately, this process should not take more than a day's effort, and should be done within a week or two of the phase's end.

Assessment of individual performances is done at the phase end review. At a minimum, individuals leaving the project should be reviewed, and depending on the length of the phase, all resources might need a review. If the review is deferred to the end of the project, the review becomes "fuzzy" due to the length of time that has elapsed.

One final task that should be done is recognition of the team's effort. Although the expressing of praise when warranted should be done throughout the project, at phase end the Project Manager should commend those who have done particularly outstanding work and convey this to senior IS management.

15.2   PROJECT MANAGEMENT REVIEW

Each project should have a Project Director who works with the Project Manager closely. The Project Director provides mentoring, guidance and monitoring of the Project Manager's efforts. The Project Director constantly reviews the project to ensure that proper Project Management principles are being followed.

It is assumed that the Project Director has been part of the planning process and has reviewed and approved the Project Plan ensuring that it meets the standards and follows the methodology.

The Project Management review involves determining if the following are being used and used correctly during the execution of the project.

- Decision Requests
- Change Requests
- Issue Logs
- Project Tracking tools
- Status Reports

- Proper Signoffs are obtained

- Meetings are held regularly (team, User and Steering Committee)

The above components ensure that proper procedures are being followed to track and control the project. But there are other facets that the Project Director needs to review:

- Quality deliverables are produced

- User satisfaction exists

- Team morale is good

15.3   TECHNICAL REVIEW

On some large projects where there is high utilization of new technology, and therefore, higher risk, independent technical reviews are appropriate. These reviews would be conducted by a senior technical resource, typically at the Technical Architect level, with the goal of ensuring that the there are no flaws in the architecture or the implementation of the architecture, that solid consistent development standards are in place and consistently used, and that the implementation of the solution is proceeding smoothly.

The reviewer inspects the overall technical architecture. He then reviews the progress of each component, the risks, and the concerns. In addition, he should review adherence to the project architecture not only at a high level but also at the code level. The review may involve:

- Technical deliverables review

- Design constructs review, including component, program, and inter/intra systems communication

- Programming standards and techniques

- Data Base Design

The intent of the review is not to change or alter the design, unless it is absolutely required. A one or two page report of the findings and recommendations is produced and delivered to the Project Manager and Project Director.

15.4 INDEPENDENT MANAGEMENT REVIEW

On some large projects, an independent management review might take place over and above all of the previously mentioned reviews. This review would encompass all aspects of Project Management and would be conducted by someone not affiliated with the Project, and may be done by someone outside of the organization itself.

By having someone outside of the Project conduct the review, there would be no bias as to how well the Project was proceeding or why it was in its current state. The review should be a totally unbiased, factual account of the Project.

Included in such a review would be a report identifying usage or non-usage and level of compliance to the items noted in the Project Management Review, as well as the following:

- Methodology,
- Project Estimating Process,
- Project Planning Process.

16.0

PROJECT MANAGER

CHARACTERISTICS

Now that we have seen the work that is required to plan, execute and review a project, let's look at the roles and traits of a successful Project Manager.

16.1   ROLES

Throughout a project, a Project Manager is engaged in many activities. Each activity requires that he perform a role. Many times the activities require roles to be performed simultaneously. The primary roles that the Project Manager will perform during the course of the project are:

Leader

Any team needs a person in charge, but just being in charge does not make a person a leader. A real leader must lead by example. He must demonstrate his ability to eliminate roadblocks, to present team concerns and have them resolved quickly and in favor of the team (as much as possible). He should understand the business process and be able to intelligently question his team and the users to ascertain what is what. He should have the conviction to take a stand and stand by it. He should stand up and state "I am ultimately responsible and accountable for the success or failure of the system" and in so doing strive to ensure its success. In essence, he should show his ability to "get things done".

Co-coordinator

Most projects are comprised of multiple groups, organizations and departments. For instance, a typical project will have a Business Sponsor, it may have users in more than one department, it may deal with an IS department who in turn co-ordinates hardware and software acquisitions and installation with vendors, and it may deal with other departments who are responsible for wiring and preparing the work area for new machines. While the Project Manager is not managing these resources per se, he must ensure that all parties are aware of the particular project dependencies on them and ensure they get their work done in the timeframe needed.

Coach

A good Project Manager should have risen through the ranks and fully understand the software engineering approach. He should be knowledgeable in the methodology and be able to mentor team members on the approach, and the reason for the approach.

Negotiator

Throughout the course of a project, many conflicts arise that require resolution. These may be formal Change Requests, or such problems as obtaining more storage, or machine access hours. The Project Manager often must negotiate the resolution of these.

In particular, not all Change Requests are as one-sided as the team might think. The Project Manager must understand the User's point of view and not be too dogmatic in his approach. The team will understand why every Change Request is not accepted or why there are compromises if there are valid reasons. They will not be favorable to increased work without cause, and where it is perceived that the Project Manager did not adequately express the team's case. Keep in mind the goal of the team and the User should be the same - the success of the project. Negotiate hard, but fair. The best negotiations are where no one feels that they have been taken advantage of. This does not imply that all Change Requests are invalid, in fact, they all should be valid. However, the team should not feel that they have been taken advantage of. Where the User disagrees, it is important to understand when to back off and compromise and when to continue to press. Remember to be partners, not adversaries.

A few years ago when I was the Project Manager for the Los Angeles County's Countywide Warrant System, we had come to a stalemate over some change requests. There were 6 or 7 outstanding change requests that the team had generated, and that I believed we should be compensated for, and time added to the schedule. The User Coordinator disagreed, and we were at a roadblock. The team needed resolution of these issues, one way or another immediately. A meeting of myself, the User Coordinator, and our respective superiors was convened. I explained each change and the reason we believed it was a change, and the User Coordinator voiced his objections. In the end, about half were resolved in our favor and half in the users. Although I was disappointed that not all were in our favor, I was glad that the roadblock was removed. The team too, was relived, and happy that the issues were resolved. It was a win-win situation - no party felt they had been taken advantage of.

Decision Maker

A good Project Manager has to be a decision maker. To be effective, he must make decisions and make them in a timely fashion.

Some large companies partially incent their management team on the following basis, with the first case being the highest compensation and the last, the least:

1)   Making the correct decision in a short timeframe,

2)   Making the incorrect decision in a short timeframe,

3)   Making the correct decision but in a long timeframe.

Interesting isn't it. The key point is that making a timely decision is **more important** than taking forever to make the "correct" decision.

Decisions, decisions, decisions. There are many decisions that need to be made every day by everyone in all walks of life. In the data processing world, there are always decisions to be made: which DBMS to use, which personnel to assign, how to structure the team, etc. However far too often a decision is not reached in a timely fashion.

Why is it that many of these decisions are not made in a timely fashion? A valid reason may be that not enough information is available to make a correct decision. If this is true, then more research is required, but a timeframe to obtain the information needs to be allocated so that the decision can be made quickly.

There are many invalid reasons for not making a decision. Primarily, there is fear of failure. If you make a decision and it turns out to be wrong, you are at fault. Therefore people often build in, as a defense mechanism, the tactic of deferring and delaying until someone else makes the decision for them, or they can form a committee to make the decision. In either instance they are buffered should the decision turn out to be wrong.

The problem with delaying and deferring a decision is that it costs time and money to not have a decision. When a project manager has 30 analysts and programmers waiting for a decision, the lack of a decision inevitably results in lost productivity, poor morale and lack of faith in management.

I was on a plane with a retired executive from a major U.S. energy company. We were discussing the problem of lack of timely decision making. He talked about an instance when he had just been promoted into the executive ranks and was in charge of all plants. A new plant was on the drawing board, and was supported by the other executives. He reviewed the requirements, the costs and the benefits and recommended against building the plant. Why would someone do this? To rock the boat? To establish a name for himself? He could have accepted the recommendation that was on the table, and whether it succeeded or not, he was not going to be faulted. However, he recommended against it because he felt it was wrong for the company. He was there for the long haul and wanted to ensure the future of the company. He made a timely decision and a very tough one. As it turns out they accepted his recommendation, and even though it proved to be the correct one, a few of the old guard were not pleased.

If you are in a position to make a decision, you certainly do not want to act hastily. It is best to obtain all relevant information as quickly as possible, and request opinions from your staff. Then make the decision. Some people may disagree with the decision, but many will be happy that it was made.

The delivery of the decision is important as well. If you have solicited input, then you should explain and justify your choice in light of the other alternatives. This becomes a little more difficult when informing your peers and superiors of the decision, especially if they disagree with your decision. Diplomacy is required in order to have your decision supported and to remain a "team player". How long do you delay taking appropriate action (to the detriment of team progress) while trying to convince your superiors of the merit of your decision?

Finally, once you have made a decision you should stick by it UNLESS new relevant information appears. This information could be contrary to the information that you used to base your decision, or it could be totally new information. It is not only appropriate, but also necessary, at this time to reevaluate your decision. If your re-evaluation results in a different decision, then you MUST admit that your first decision is no longer valid, and that due to new information being available, another decision has been made. This isn't being flippant or wishy-washy, simply intelligent. People will appreciate your ability to change a decision based on new input. Do not be stubborn and stick with a decision simply because it was right at the time and you do not want to rethink it. As tough as the original decision was, you owe it to yourself and your company to reevaluate the decision.

16.2   TRAITS

In addition to performing the roles above, the ideal Project Manager should possess the following traits:

A Good listener

It is important to really listen to people. Many of you I'm sure have talked to someone who clearly is more interested in ending the conversation than listening and understanding what you are saying. People can sense this (it's even easier when the individual taps his foot, or begins to read). Being a good listener means being attentive, and trying to understand what the person is saying. If you do this, it not only conveys respect for the individual but provides you with the opportunity to suggest some action or agree on some action after careful consideration

Receptive to Suggestions

I always make a point of trying to use other people's suggestions. Whether in analysis where they have a different approach, or just a suggestion to increase team morale, if the suggestion is good I try and use it. Even if the suggestion is slightly poorer than mine

I still try and use it. Why? Because it fosters a creative spirit in the individual. He knows that if he thinks of a better way of doing something it may be accepted, and not turned down without consideration. People respond by being creative and thinking, which is what we want from all people on the team.

Receptive to Learning

Throughout the project, a good Project Manager should be able to learn constantly. It may be some of the application, some of the technology, how to utilize project tracking tools better, how to use project management principles, etc. As long as you hope and expect to learn something you will.

Integrity

Just as you the Project Manager expect your staff to report accurate status to you, you too must report accurately to your management. In doing so you provide them with the ability to help you. It is also important to recognize the "right" way of doing things and if necessary, take a step backwards to rectify a problem rather than attempt to "fix it" later and hope no one notices. The goal should be for the project to be a success – not for any personal gain.

A Sense of Humor

Certainly projects are to be taken very seriously. People's careers, the company's profitability, and other important things may depend on the success of a project. You must do your professional best to accomplish these goals. However, some humor and fun will increase the probability of success. If you celebrate phase completion, have a baseball game with your end users, or whatever, you are fostering comradery. The team that plays hard works hard. People should enjoy coming to work and look forward to occasional team celebrations.

Available

If you are the Project Manager you ought to be available for questions, guidance and resolution of issues. If people can't find you or you have no time for them, they will stop bringing things to your attention. When they do, you can be sure that disaster is right around the corner.

Communication Skills

As the Project Manager, you are expected to provide written status reports, and be able to provide oral presentations to IS and Users. Effective speaking and writing skills are a must.

Commitment to Excellence

Every project with which I am associated I want to be the best in the organization's history. As the Project Manager, I am proud to have my name associated with the Project. The Project Manager should be proud of his system, and do everything in his power to make it an unqualified success.

Technically Knowledgeable

No one expects a Project Manager to be a technical whiz. However, many do not even possess the basic knowledge of the technical components of the systems they are managing. At a bare minimum, every Project Manager should understand **what** software, hardware and communications are used in the system. This typically involves reviewing the following documents: the "Technical Alternatives", the "Technical Requirements" document and the "Technical Specifications". One trick I've learned is that if you can't understand something because it is new to you, have someone (preferably your Technical Architect) explain it in terms you are familiar with. In that way you have a frame of reference - "Oh its like xxx in a mainframe environment".

Focused

As the Project Manager, your first goal and objective has to be to ensure the project is a success. You should not waiver from this. This may mean such things as delaying education, not participating in reviews, and staying with your current project until it is successfully implemented. It is important that you stay focused on doing everything you can to ensure the success of the project. I have even delayed vacations when I perceived my absence would jeopardize the success of the project.

Personable

People must not be intimidated by you. If they are they will not come into your office or cubicle to ask questions or inform you of problems. Encourage people to come in and when they do, give them your attention. Treat them with respect and never, never admonish people in public.

Be the Heavy

The success of the project is the Project Manager's responsibility. Because of this, he is responsible for ensuring that all people on the project are performing to the best of their abilities. He may have to take some actions with some individuals to improve their performance, or may have to replace them on the project. Individual personalities and relationships should not interfere with the best judgment of the Project Manager.

An Amateur Psychiatrist

On larger projects there may be many personality clashes. Each of the individuals may be very good, but problems in working together interfere with results. The Project Manager needs to talk with each individual about putting aside personality problems and working with the rest of the team. He may have to re-assign personnel to different areas of the project to avoid the direct day-to-day interaction of the affected parties. In a worst case scenario, one or more of the personnel may have to be replaced on the project. The Project Manager should try to avoid the latter, but if the Project is in jeopardy he must take appropriate action.

Understand the Methodology Implicitly

The Project Manager has to be an expert in the Methodology and Project Management Principles. He should be able to explain why the project plan was constructed the way it was to the team and management. He should mentor the people on his project that are aspiring to project management responsibilities so they learn the proper approaches, and *the reasons behind them.* The response "the methodology says so" is not the productive answer. The Project Manager needs to know and understand when modifications to the methodology are allowed and be able to rationalize these changes.

Trust the Team

The Project Manager should trust the team's input and carry their concerns, problems and status to management. Certainly the Project Manager needs to understand the team's view, and if he agrees, take appropriate action; for example, for a particular problem some specialized hardware or software is required. This may or may not exist in the company, but if it is really required, the Project Manager should recommend its acquisition.

Praise the Team

Too often the only time Project Managers talk with team members individually is when there is a problem. Maybe the person is falling behind, or there is a problem with quality. It is just as important, if not more important to praise team members when they have done a job well. People respond to praise that is warranted. They perceive that people recognize a job well done and are happy and proud of the recognition. Just think how you feel when your boss praises you!

Make Decisions

The Project Manager needs to keep the project moving. Often times there are delays in making decisions that hamper the project team's ability to move forward. The Project Manager must anticipate problem areas, create decision requests, and / or hold special meetings of the Steering Committee (if necessary) in order to make sure a decision is made in a timely manner. The Project Manager should fully understand the issue and options and help guide the decision by providing options and recommendations. They cannot be indecisive and foster or further the delay in making decisions.

Figure 16.1

ALIGNING PROJECTS WITH ORGANIZATIONAL FRAMEWORK

17.1 OVERVIEW

So far, this book has addressed how to estimate and manage projects that work within an existing framework. But what if the organization is changing its business and technology views? What is the proper sequencing of steps to re-organize a business and change technical platforms? When can IS successfully develop systems that leverage the new business look? This chapter addresses these issues.

While this document does not purport to be a technical reference guide, it is important to understand some of the technology components, how they intertwine, and when new projects can begin to use the results of these components. The components that I refer to are: the Enterprise Wide Architecture (EWA), Technology Selection and Deployment (TSD), System Development Environments (SDE), and Business Process Re-Engineering (BPR). It is important to understand these components, and determine where your organization is relative to these components *before* embarking on any new technique or methodology.

17.2 ENTERPRISE WIDE ARCHITECTURE (EWA)

What is an Enterprise Wide Architecture and what is its purpose? Its scope is the enterprise (or in some cases, a strategic business unit). An EWA defines the standard infrastructure to provide computer services to meet the business needs of the entire enterprise. The purpose of an EWA is to define a standard environment for the enterprise. Reducing the diversity of technology present within an organization provides significant savings in costs of acquisition, support, development and migration. Risk and development time are reduced through the reuse of proven components. Developers do not spend time in non-productive technical assessments and continual re-training.

What are the steps and what exactly is the output?

Typically there are four major steps:

- Assess Current Situation

- Define the Architectural Principles

- Define the Enterprise Requirements

- Develop Architecture Strategy

Assess Current Situation

The objective of this step is to assess how well the current situation supports or doesn't support the business objectives. The current situation is documented in terms of business requirements, availability, security, ease of use, cost of operation, response time, maintainability, performance, technology and interfaces. By technology I mean what languages the systems are written in, what operating system is used, what data base(s) are used, degree of distribution, what topology is being used - mainframe, Client server, Internet, etc..

Define the Architectural Principles

Architectural principles are defined that address the objectives of IS management in terms of technology Architectures and deployment, systems development and maintenance. All future development should follow these principles, or if it does not, there needs to be a rationale (grounded in business terms) for why not.

Define the Enterprise Requirements

The requirements for new systems are also documented but with less emphasis on the technology and more on what is required from technology (i.e. instead of specifying a Wide Area Network (WAN), state the requirement for data to be shared at 3 remote sites across the country.)

Develop Architecture Strategy

The deliverable is a high level definition of the architectural components required by the business. Architectural components include organizational, process, technical, application and information.

For the technical architecture, think in terms of a diagram that depicts all of the hardware, software and communications requirements without being product specific. You might see a mainframe connected to a Local Area Network (LAN) supporting multiple workstations with an indication of line speed requirements. You would not see all of the bridges, routers, etc or the names and exact configurations of any of the pieces. These come later in the Technology Selection and Deployment (TSD).

The goal of the EWA is to produce a stable architecture that provides for today's requirements and for long term growth. The exact technology deployed to support the architecture may change, but the architecture remains much more stable. For instance, the architecture may state a need for a distributed relational data base. The TSD may have initially selected Oracle to handle this requirement. Over the course of time, other Data Base products may replace Oracle (DB2, SQL, etc), but the underlying requirement for a distributed relational data base has not changed.

17.3 TECHNOLOGY SELECTION AND DEPLOYMENT (TSD)

The Technology Selection and Deployment phase is intended to define the Architecture components as actual technical components. For instance, it takes the architectural statement of the need for a distributed relational database and defines which products to use and what interface pieces are required.

You may think it is simple to select the best, based on what different trade organizations such as Gartner Group and Forrester say are the best. However, as Pat Smith former Senior Vice President Systems Integration and Technology Services of Sprint states, "There will be ongoing temptation and pressure to select only the "best products". However, the classification of "best product in the market", as evaluated in the narrow perspective of its features versus those of other products in a category, is irrelevant for a particular organization"[6] . What

---

[6]Patrick Smith, *Client/Server Computing*, SAMS Publishing, Indiana, 1992, pp. 168

is relevant is to select architectural components that work well together, for the features that have importance to the organization.

After selecting the appropriate new technologies, and the interfaces required to converse with the existing platform (assuming that this is a requirement), the next step is to deploy the technology.

Do you acquire all of the pieces now and implement the complete architecture? Unless it is really simple and straightforward, the answer is no. Instead the business requirements driving this process are reviewed. What are the most pressing business requirements and what do they require from the technology? Assembling only components that are required makes sense since the cost of hardware continually drops. Deferring the buying of 100 workstations until they are absolutely needed will save money for your company. Only those pieces required for the application should be acquired. However this does include the components that are required by the Systems Development Environment (SDE) which is used to build new systems (more on the SDE in the next section).

17.4 SYSTEMS DEVELOPMENT ENVIRONMENTS (SDE).

A good definition of an SDE is given by Peter Somers, former Chief Technologist of Indicom Technology, "A SDE is an environment consisting of tools and associated processes that allows programmers to create new system code faster and with better quality, while maintenance programmers can correct bugs and enhance systems more efficiently."[7]

Every project should be constructed using a SDE. It is up to each organization to look at the suite of tools they intend to use for development and maintenance and acquire or create a SDE that more effectively leverages these tools. Keep in mind that part of the SDE is standards and procedures that instruct developers to use the tools optimally and provide techniques and tips. The standards and procedures should also address how the analysts will use the SDE. This will help the analysts design towards the environment rather than away from it. This requirement means that the SDE development must occur early in the project.

Note that there may be several SDE's in place in an organization; one or more that have been used to support existing systems development and maintenance, and one for the new systems development.

---

[7]Peter Somers, former Chief Technologist, Indicom Technology, personal communication

17.5   BUSINESS PROCESS RE-ENGINEERING (BPR)

Business Process Re-engineering (BPR) has been used to define re-engineering of entire or substantially the entire enterprise as well as re-engineering a specific business process.   The later has always been a function that proper analysis should have accomplished.   There are many books and techniques for BPR. The reason for addressing BPR in this book is to define when it should take place relative to the other technical components. For this purpose, we will use the definition of BPR that addresses the entire enterprise (or substantially the entire enterprise), and not the re-engineering of a specific business process (where the EWA should be in place already).

17.6   ORDER OF PRECEDENCE

How should the projects for EWA, TSD, SDE, BPR and the application projects themselves be deployed in order to be successful?   Figure 17-1 depicts the various components and a timeline for when they should occur. There are two phases to all of the components, the first phase produces deliverable(s) which define the desired outcome (architecture, business lines, technology, etc), and the second part is the implementation of these.   The first box for each component is the deliverable and the second is the implementation.

As you can see, the BPR deliverable should be done first, and should be mostly completed prior to the EWA.   The BPR process removes from consideration how business is done today, where it is done, and how technology is used today. The new business state will provide the requirements for the EWA. If you have constructed the EWA and then re-engineer functions, your EWA may be obsolete. Having said that, it is important during the BPR function to look for potential uses of emerging technologies.   Without looking at these, the re-engineering may not produce the optimal outcome. The technology should not however, drive the process.

The latter part of the BPR deliverable and the early part of the EWA where we are gathering information about existing systems may be done in parallel. But when we start to look at requirements for the future, the BPR deliverable should be complete since many requirements will stem from this endeavor.

The Technology Selection and Deployment (TSD) generally starts after the EWA. However, if parts of the EWA are rock solid and projects are waiting, the TSD can begin before the EWA is complete.

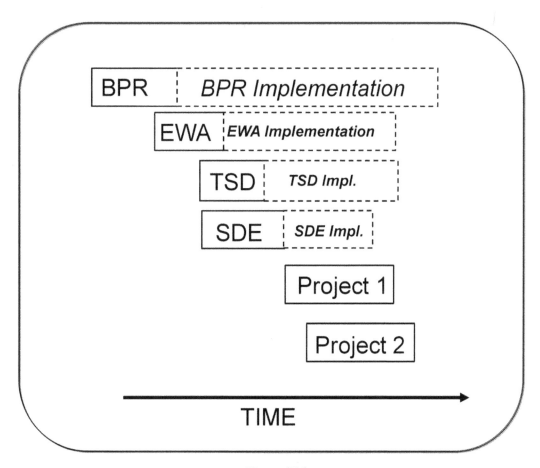

Figure 17-1

During the latter part of the EWA, some components of the SDE may begin. These are components that are not associated with the technology, such as methodology and to some degree, standards.

All of the implementations of BPR, EWA, TSD and SDE may be running in parallel, driven by business requirements, budgets and resources.

The first projects that utilize the new architecture and technology can begin when the SDE is ready for them (the SDE does not have to be complete) and when the architecture and technology required for the project has been deployed. Again, the projects if they are in General Design may be begun earlier, but as soon as the technology issues arise in Detail Design, the architecture and technology should be selected. It is prudent to not try to complete General Design until the architecture and technology are ready, in order to avoid any phase gaps.

18.0

DEVELOPMENT

METHODOLOGIES

AND PROJECTS

18.1 OVERVIEW

So far, this book has addressed how to estimate and manage projects that are more traditional, and to some extent supportive of a "waterfall" lifecycle. Although the waterfall lifecycle has been around for quite some time, most organizations continue to use this, as it is most widely understood, and quite structured. Not too many organizations have gone to the Agile methodology (a version of rapid application development) for many reasons including fewer experts and fewer people having worked with this method of developing systems.

Each of these methods: Prototyping, Rapid Application Development (RAD), and Object Oriented (OO) Development, has merit, and can provide many benefits if used correctly. Unfortunately, many organizations eager for improvement jump on the latest bandwagon and expect the panacea that never occurs.

Why is this? Why do many organizations fail to realize the benefits of new technologies and methodologies? Usually for the same reason that they fail in the older techniques - poor planning and poor execution of the plan. You must really understand the new technology in order to create achievable plans. No technique, methodology, tool or language is going to help IS people to develop better systems faster if they do not understand the fundamentals of what they are trying to accomplish.

The great successes of all these techniques are prominently described in all the computer magazines. When an end user reads these, often there becomes intense pressure to "get with it" and develop new systems as fast as other organizations in order to become more competitive or to maintain the competitive edge. However, most of the competition is failing at these endeavors. The ones that succeed are the ones who have taken the time and effort to understand the techniques, understand how they differ from the current approach, and understand how they ought to fit into the organization's methodology.

Knowing when and how to use a technique is absolutely imperative. It is only with this knowledge that a proper plan can be crafted.

18.2 SOLUTIONS AND REQUIREMENTS

Rapid Application Development (RAD), Prototyping, and Object Oriented (OO) development are all valid and excellent techniques when used in the right fashion and with the right environment.

However some people believe that these techniques (or some of them) obviate the need for proper analysis. People have complained that the techniques and methodologies used by analysts don't define the user's requirements adequately or in understandable terms. In my opinion, they do. The problem is with the analysts who use the techniques and methodologies. Many are never trained correctly, and many are not adept at "ferreting out" the user's true requirements. Let's train and mentor our analysts properly on whatever technique/methodology we are going to use and *then* look at the results.

There are Users who desire to dictate functional and technical solutions to their problems before the requirements are determined. What is so wrong with their dictating the functional solution? Depending on the individual, everything. Solutions should not be determined until the requirements are known. When all requirements are known (at a point in time), a solution may be crafted that addresses the requirements in an optimal fashion.

For instance, a large County had a requirement for a "Warrant" system. They hired a company to do a project definition. Well, "warrant" was generic for eight variants of warrants - wants, Temporary Restraining Orders (TROs), etc. Even though 9/10's of the data and processing was the same for these variants, the design had eight different add screens, eight different change screens, eight different "recall" screens etc. This made the system appear more complex than it was, because there were many Menus to get to the right spot, and obviously many many screens. A more logical approach in looking at the system's requirements was to allow all adds from one screen (partitioning off unlike data) and to control the input by verifying the type of warrant and ensuring all data pertinent to that type was entered. A lot fewer screens, a lot simpler for the operator.

What's wrong with a User dictated technical solution? Hopefully the IS department is more knowledgeable of technology than the User, and, hopefully the IS organization has an enterprise architecture and technology deployment that all departments adhere to. For example, have you ever tried to share a spreadsheet or word processing document with someone else in your company only to find out they use a different product?

Mortgage Example

It is easy for someone to calculate a mortgage over the term of a loan, given an interest rate and the term of the loan on his own PC using Excel (I did it for my mortgage). If a company was in the business of loan administrating and part of the job was to handle inquires on how much a loan would cost over the course of the term, and they were handling only a few mortgages, then this seems a simple and cheap method. This illustrates the faulty thinking of many users. They have "programmed" something at home on their PC and expect it to be as simple at work. We probably all have heard, "Why does it take so long? I did ... on my home computer in only half a day!".

Let's return to our mortgage example. As it turns out, many of the people who inquire at a mortgage company request a print out of the loan calculation before accepting the loan. Now someone needs to take down their name and address for mailing. It turns out that the sales department would like to follow up on those who have inquired about but not yet taken out a loan. Where is the information? On a slip of paper? Thrown away? On the top of the spreadsheet?

How can Marketing have access to the name, address and mortgage calculation for follow-up?

What happens when the client says OK? Now the Loan Origination department wants the current address, phone number etc. (You don't want to ask the client again do you?).

To make it more complex the Loan Origination department puts the loan on a mainframe system, and the Marketing and Loan Inquiry departments are on different LANs. There is more than one Loan Inquiry representative and the rates can change every day. Who updates each person's personal copy of the spreadsheet?

Enough! Knowing why, where and when each person needs which information and being aware of the various required interfaces must come before deciding on the implementation architecture (EXCEL on a standalone PC). Understand the implications of a solution before deciding it is the correct one. And to do so, you need to define the requirements first. In the above example, if the requirements for data sharing were known, the correct solution could have been crafted and deployed.

My friend the expert said ...

Recently an IS person told me that a new system was going to use FoxPro as the Database; even before the requirements were determined. I asked him why. His response was that the head of the department's brother in-law had used FoxPro and told him it was an excellent data base! Who knows whether his brother-in-law is a data base expert or a plumber?

Users should never dictate the technical solution. They may offer a solution that IS can review, but IS is responsible for ensuring the technical viability of the project, and the technologies compliance with the EWA.

There are also technical personnel who also like to state solutions without knowing the requirements. This too is the wrong approach.

Define the requirements, understand the requirements and their implications, and then determine the solution.

Finally, make sure that there is a workflow analysis – this is part of proper analysis, but seems to have been forgotten. Recently in the implementation of an EMR system at a hospital, the manual dictation of Physician notes was being replaced by a system where the Physician entered the data on to a screen in the EMR. Sounds simple and sounds like a good solution (the dictation system could have 24-48 hour turnaround where with the new system it was immediate). However, the current system workflow was not analyzed to see the downstream effect on billing. As it turned out, in the old solution, the dictated notes were sent to billing. In the new solution, no one realized this link would be broken. The result was some of the Physician billing was not done for a few months, and once it was discovered, a patchwork Excel spreadsheet was required to send to Billing. If the proper analysis had taken place, *before* implementing the solution it would have uncovered this problem.

18.3 PROTOTYPING

A prototype implements components of a system and presents views of how the system will look and interact with the user. The prototype varies as to how much of the final system is built, it can show simply the screen and report layout, or implement increased layers of edits, database access, logic, security, and navigation depending on the completeness required to gain user buy in.

In the simplest sense, when we drew screen layouts and used the old paper report form to depict how a report would look, these were "prototypes".

Let's differentiate between two specific types of prototypes: Horizontal and Vertical. View the system as a large rectangle, as depicted in Figure 18.1. Each system is composed of many externals (screens, reports and "other processes"). Now list all of these externals horizontally at top of the rectangle. This represents all of the processes in the system that need to be analyzed, designed, coded and tested. As you analyze, design, code and test these externals, you progress vertically in creating the complete system.

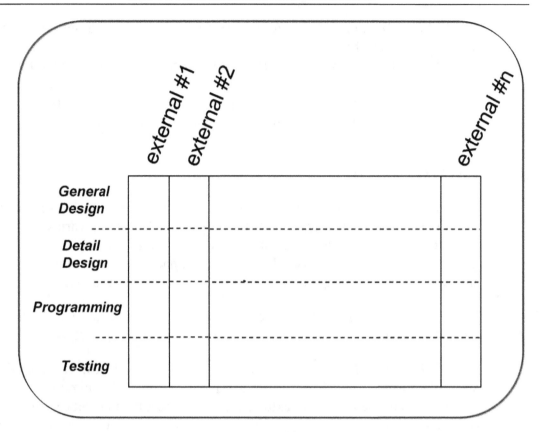

Figure 18.1

A horizontal prototype consists of all screens and reports that comprise the total system. Typically the "other processes" are ones that the User is not aware of, or are not pertinent to him, and may not be able to be depicted visually. Many times the tool used to create the prototype is not the same as the tool used to actually develop the system. If the tools are different, extreme care must be taken to ensure the prototype does not describe something that is very difficult, if not impossible, to build in the development tool. Horizontal prototyping involves creating samples of all screens and reports across the horizontal access of the system rectangle depicted in figure 18.1, and is depicted in figure 18.2, on the next page.

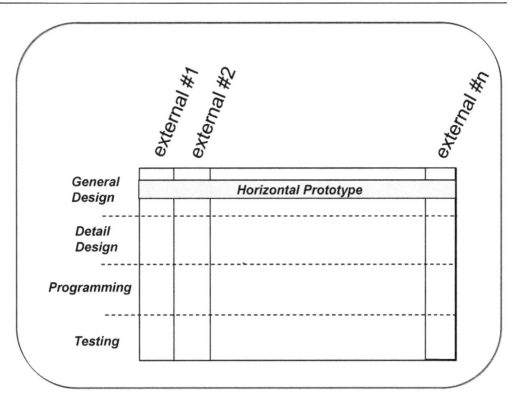

Figure 18.2

A vertical prototype is used when one or more of the externals is analyzed, designed, coded and tested (either in whole or in part with stubs). To complete the vertical prototype, some of the technical processes may have to be written and test files created. The vertical prototype is sometimes referred to as a Proof of Concept as it may be used as proof of the technical viability of an approach before the project is committed to the approach. For instance, in General Design, the vertical prototype may test the end to end communications in a client server environment (if this is the first implementation). The test may demonstrate that some of the components do not work together as expected (or documented by the vendor). New solutions can be researched and tested before the application is committed to the approach. The vertical prototype construct is depicted in figure 18.3.

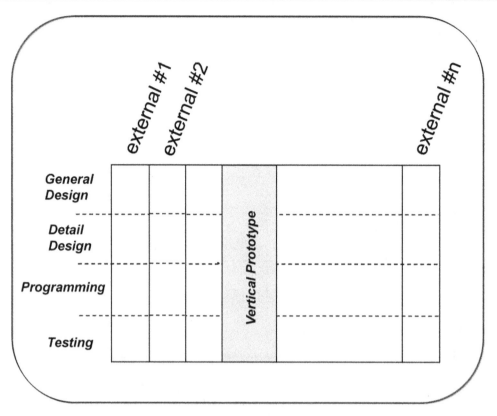

Figure 18.3

In traditional, client server and some OO approaches, a horizontal prototype is constructed in the General Design Phase and used to define the Scope of the system with the User. Vertical prototyping is used in traditional, client server, and OO development to prove an approach or to prove a design component that is critical to the overall development. This is typically done to either verify the approach or to develop some of the more complex code and demonstrate its viability prior to embarking on a design that is dependent on it.

Prototyping is used in RAD somewhat differently. In a RAD approach, successive iterations of the prototype are constructed, reviewed, and modified by a team of IS and user personnel. In the end, the final system is a refinement of the initial prototype.

There may be some use of prototyping in BPR to demonstrate or validate a concept, but Prototyping is not fundamental to BPR.

In order to do prototyping, you need a SDE in place that has the tools to support the intended level of prototyping. Hopefully all new systems will utilize the SDE based on the technology developed from the Enterprise Wide Architecture, and not the old SDEs for existing systems.

18.4 RAPID APPLICATION DEVELOPMENT (RAD)

RAD uses Prototyping to rapidly create iterations of a system. The RAD process usually begins after Project Definition (a problem we will discuss further), or may start after General Design. A small team composed of business resources, analysts, and programmers and a Project Manager stays together for the duration of the project. In traditional development methodologies the User is heavily involved in analysis, acceptance testing and implementation activities and very little during design and programming. In RAD, they are actively involved through implementation. The RAD team creates versions (or iterations) of the system quite rapidly using horizontal and vertical prototyping. The prototyping tool must be a robust tool that can generate efficient production quality code.

The team creates screens and reports and other programs to support them, reviews these and makes refinements, then creates some more, makes more refinements etc. In some situations, a "stable" subset of the project may be staged in a central repository and as new additions arise, they are added to the "stable" version to create a new stable version. This continues until the system is complete.

When should you use RAD and how? RAD works best on small projects in the 4-6 month range when the system is a standalone system with a minimum of interfaces to existing systems or external systems. The short timeframes for development mean that interfaces to systems over which the RAD team has little control will significantly increase the risk of schedule slippage.

RAD development challenges the Project Manager because there is a tendency to add more functionality during the process. For the purposes of establishing a baseline for the initial plan, specific estimates of the number and type of externals should be made at the outset. A contingency factor should be determined to allow some growth but yet provide parameters for the growth.

For RAD to work *well* in a *non-standalone* environment (which describes most systems), there must be an Enterprise Wide Architecture (EWA), the technology must be deployed, and there must be a robust System Development Environment

(SDE) that has clear standards and tools to help the developers. Even more important it should contain the service level functions of: data base access, security, screen management, communications to other Data Bases and other systems. Then the RAD team can concentrate on the application and can develop the screens and reports quickly with the needed assurance that the system does fit into the existing enterprise structure.

Too often RAD projects become bogged down creating programs to provide these services, or worse they do not, thus creating a standalone system that ought not to be standalone. The result is a lot of re-work to integrate the system into the enterprise. This is the situation encountered frequently in systems developed by niche client server development houses working directly for an end-user department on a specific business application need.

It is also important if the system to be developed is large, and components are to be developed independently, that the base functionality is determined and built first. Figure 18.4 depicts an overall project and the poor partitioning of the system into 4 separate projects. Each of the projects has some portion of the base functionality but not all. If one were to develop this system in this manner, it would likely be a failure.

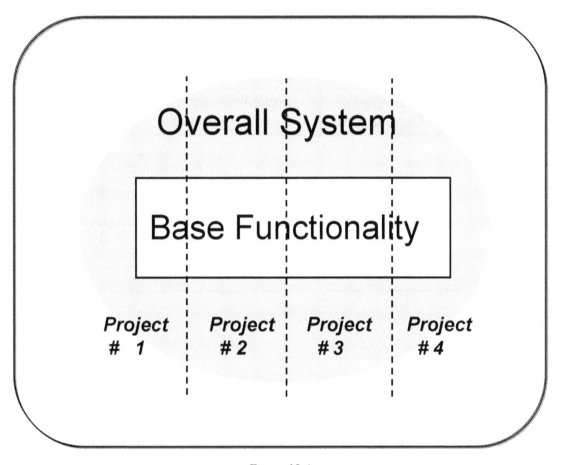

Figure 18.4

Figure 18.5 shows a more appropriate method of attacking the problem. First, "all" of the known functionality for all of the system is defined at a point in time. This includes the base functionality and all other known requirements. From this, the system can be partitioned into sub-projects, one of which is to implement the base functionality (and perhaps some additional functionality) and the remaining partitions based on the business benefits to implementing the remaining components. Each project should be less than one year. In this way, the system is developed logically without the potential for a large amount of re-work based on future uncovering of 'base' requirements.

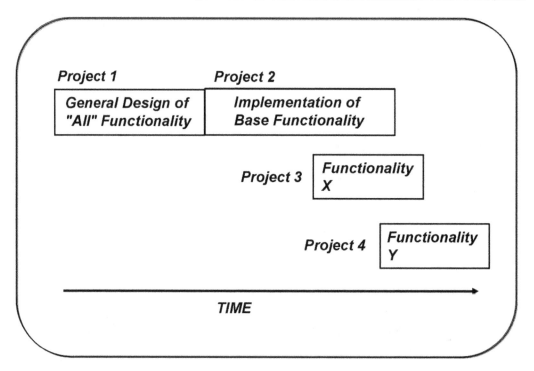

Figure 18.5

It should be noted that the approach above is the same for RAD and any other method - analyze the base first, develop the base, and then develop sub-components. As indicated, there can be some overlap of these projects as well.

RAD projects are risky because most often the SDE is incomplete, the scope increases, the interfaces to other systems are not designed into the system and the result may be a standalone system that takes longer to complete and has limited functionality.

However the approach can and does work quite effectively in the right situation and with the right disciplines and skills on the team. RAD is essential for organizations under the gun to bring new products to market rapidly. Success will occur when the RAD discipline is implemented in the team. The demand for RAD leads to an interest in Object Oriented development methodologies and tools.

18.5 OBJECT ORIENTED (OO) DEVELOPMENT

18.5.1 Overview

Object Oriented (OO) development has been around since 1967 with Simula '67. However, it is just in the past 10-15 years that many companies started pilot projects using OO, and only in the past 6-10 years that major corporations have turned to OO development for large scale development rather than "tests". Although most companies now use OO for some or all of their projects, many have yet to make that move.

Many of the OO Methodologies stress that the initial development is just one small component of the overall life cycle, which includes the maintenance life. When viewed in light of the overall life of a system, the initial development costs are but 10-20%. While this is true of all systems regardless of how they were implemented in the past, the objective of OO development is to produce a more robust system that is easier to change thereby reducing maintenance costs.

OO development proposes to achieve lower maintenance costs by having objects that can be used over and over again where they are required, and by such concepts as inheritance, whereby similar objects need not be completely redeveloped and tested, but can inherit traits and characteristics from other objects.

18.5.2 Requirements and OO

Many of the OO methodologies will indicate that it is impossible or virtually impossible to state all of the requirements of a system before beginning development. This is true and is not different from other methodologies. It is important in all methodologies to define as many of the requirements as possible, and to control changes via change control.

In OO methodologies, there are strong urgings to develop the system in iterative steps. Again, this is no different than what has been stated earlier in this document - that it is best to have small projects, less than a year in duration.

Where some OO methodologies seem to go "off-track" is in their leading users to assume that you should begin to develop the "system" in small pieces, implement them into the workforce and then begin on the next piece (or have a number of pieces ongoing at one time). While this is feasible for small standalone systems, for large integrated systems the base functionality needs to be defined *before* developing the supporting pieces. Ivar Jacobson states "For sequencing of development stages to be successful, it is essential to define stages which do not necessitate changing the results of earlier stages as the later stages are introduced. Thus it is important to capture early those requirements which form the base of the entire system"[8]. Unfortunately, it is not clear what is base functionality and what is not. By conducting a General Design, all of the requirements at *a point in time* will be documented. Analysis/paralysis should be avoided. The known requirements can then be partitioned into separate sub-projects for design and implementation, with the base functionality system being the first project. This approach is obviously not unique to OO development but rather, once again, is good software engineering that can be deployed in traditional methods.

One of the features of the approach defined by Jacobson and others is the ability to conduct the General Design (Jacobson uses the term Analysis) and to estimate the project to completion (as we can in traditional development). Other OO approaches which seem less useful in a business environment require the project to proceed until the end of Design before a valid estimate (and scope) can be defined. Most business users need to understand the costs and the benefits earlier and make business decisions as to whether it is cost effective before investing too much effort.

18.5.3 OO and Traditional Methodologies

Many organizations are reticent to jump into Object Oriented development in a big way. There are ways to obtain benefits from OO without having to use OO techniques throughout the lifecycle. It is possible, and sometimes advantageous to use traditional Structured Analysis approaches to General Design, and then use Object Oriented Design (OOD) to determine and design the objects. Finally after OOD it is possible to use Object Oriented Programming (OOP) or other non-OOP languages to develop the system. It is clear that OO Programming without OO Design is unreasonable, and OO Analysis without OO Design does not make a lot of sense.

I have seen successful OO Design use as a base the results of traditional structured analysis techniques. However, it is more work to go from structured analysis to OO Design than to go from OO Analysis to OO Design.

There are a number of reasons organizations may not use OO throughout the

---

[8]Ivar Jacobson, *Object-Oriented Software Engineering*, Addison-Wesley, pp. 27

project:

- One of the greatest benefits in OO is in the robust design that is developed out of OO Design, and therefore many of the benefits are achievable by using OO Design only.

- The General Design may be complete, or near complete. Since there is a steep learning curve and hence a large training effort in all facets of OO development, you may want to bite off smaller chunks of the training effort at one time.

- If you utilize all OO techniques on a project, the risk is higher and depending on the critical nature of the project, this may be unacceptable to management.

Today OO techniques are no longer considered the leading edge, but rather are stable and mainstream. They are also not a panacea. Organizations that approach OO correctly by ensuring there is an implemented EWA, an implemented SDE, and adequate training in OO for the developers, will be more successful than those who jump on the trend and expect magic without investment and effort.

With many companies jumping on the OO bandwagon, care must be taken to understand the actual OO support provided by a tool and implement the disciplines to use the tool as it is designed to be used. Unfortunately it is possible, and likely, to design and build a structured system implemented with mainframe programming language constructs while using only OO methodologies and tools. This happens regularly when traditionally trained designers and developers are instructed to use OO techniques but not provided with trained OO experts as mentors.

18.6 SUMMARY

Prototyping, RAD and OO are all methods that can and should be used to develop systems or portions of systems. In and by themselves they are not a "silver bullet". It is important to understand how they can be used with the current methods used at your organization, when they should be used by themselves, and whether they should be used at all for a particular system.

All projects should use horizontal prototyping in the General Design phase. It provides the users with a better picture of what the system will look like.

Vertical prototyping should be used when there is concern over some technical aspect of the project. It should be done as soon as possible in the lifecycle in order to prove that the concerns are invalid, or if the concerns are valid, allow enough time to determine an acceptable alternative.

RAD projects should be small (less than 6 months) for the first implementations. RAD projects are more successful when there are few interfaces to other systems and there is an environment that provides for the technical aspects of the project (communications, data base access, etc). Strong project management practices are required to ensure the projects do not have expanded scope without business reasons and management approval.

OO is here to stay. There are many benefits associated with OO, but fundamentals such as good analysis and project management are as important as they were in traditional methodologies. Proper planning and training of personnel in OO methodologies will lead to successful OO projects.

APPENDIX A:
HIGH LEVEL PLAN EXAMPLE

This appendix is intended to provide a reasonable real life example of how to take the programming estimate and develop a high level Gantt and a low level Gantt. Along the way, estimates are tuned and validated. At the end of the process, a reasonable high level plan is created that details the duration and man days required.

A.1 DEVELOPING THE HIGH LEVEL PLAN

From the estimates that were developed using one of the methods described in chapter 5, you have an estimate for the overall programming estimate. You now want to use this estimate to get overall estimates for all phases in the life cycle. From this, you can plan the overall phase durations using a Gantt chart. Finally, using the high level Gantt chart and the overall estimates, you want to map out each team member's involvement in each phase on the low level Gantt to ensure you have the correct number of resources and have not under-utilized or over-utilized any one resource. This is accomplished by doing the following:

1) Generate man days for all phases

2) Review deliverables required

3) Create High Level Gantt (phase level)

4) Tune the High Level Gantt

5) Create Work Plan for First Phase

6) Create Low Level Gantt (phase level)

7) Tune the Low Level Gantt

After this is done, you should have an overall duration and an overall estimate for the project that you can take to management and believe that you can achieve it. Let's go through an example of these 7 stages.

A.2 Generate Man Days for All Phases

For our example, let's assume there were 100 externals and that the programming effort worked out to be 832 man days. But, of those 832 days, 32 were for technical "other" programs that do not need all phases. Subtract the 32 days from the 832, leaving 800. This is the total programming effort. Now apply the standard percentages by multiplying the percentage for each phase times the programming man days (800). Spreadsheet A.1 shows the phases and the standard percentage for each, and the man days arrived at by doing the multiplication.

Conversion of 800 Programming days to overall Project Days		
PHASE	**Percent**	**Man days**
General Design	22	176
Detailed Design	44	352
Programming	100	800
System Test	34	272
Acceptance Test	22	176
Manuals	16	128
Training	16	128
Project Management	34	272
Technical Architect	20	160
Conversion	0	0
Facilities Install	0	0
Total		2,464

Spreadsheet A.1

Now, take the 32 "other" programming days and for each phase use the percentages multiplied times the programming days of 32. Spreadsheet A.2, column 4 shows the result. Now, adjust the technical days by using only the man days for each phase that is appropriate. For the purposes of our example, General Design, Detailed Design, Programming, System and Acceptance Testing are required for the technical programming. This is depicted in column 5 of spreadsheet A.2. Finally add the technical days that we expect to use (column 5) to the estimates for the overall project (column 2). The results are in column 6 of spreadsheet A.2.

	Programming man days = 800 "technical" programming days = 32				
PHASE	Percent	Man days for 800	Man days for 32	technical days used	Total Project
General Design	22	176	7	7	183
Detailed Design	44	352	14	14	366
Programming	100	800	32	32	832
System Test	34	272	11	11	283
Acceptance Test	22	176	7	7	183
Manuals	16	128	5		128
Training	16	128	5		128
Project Management	34	272	11		272
Technical Architect	20	160	6		160
Conversion	0	0	0		0
Facilities Install	0	0	0		0
Total		2,464	99	71	2,535

Spreadsheet A.2

There is an alternative to eliminating the 32 man days of technical programming and determining the phase man days and adding in the additional days for technical programming across the phases where it is appropriate. You could leave the total programming days at 832, calculate the phase man days and then eliminate the days from the phases that are not affected by the technical programming. Either way, the numbers come out the same, so it is a matter of preference.

Now, we want to review the overall project and decide if all of the phases are required for this project. In our example, let's say that there is no Conversion.

For Facilities Install, we need to set up a small 4 workstation LAN for development and a 10 workstation LAN for production. We requested an estimate from the technical services and they indicated that for acquiring, installing and testing the LAN for development would be 30 man days and take

4 weeks elapsed time, while for production it would take 20 man days and take 2 weeks in duration. We will need to depict the durations on our Gantt, so we need to keep them around, but for now, we need only add 50 days to the Facilities Install time on our spreadsheet.

We expect the user department to conduct the training and develop the User Manuals (reference material), and other than that time, there is no other Training or Manuals time required from IS personnel. For the User Training and Manuals, we need to determine the amount of time that will be required by the User from the Information Systems (IS) personnel in order to accomplish these tasks.

For Training, the User will need someone familiar with the system to answer questions, and provide the trainers with some training. Also, someone will need to set up a training environment for them and document how the User can manage it (specifically, how do they reset the system back to its original data contents at the beginning of each training session?).

Although the User is writing the User Manuals, someone knowledgeable from IS will be required to provide answers and guidance. In looking at the overall estimates, there are 128 man days effort for each of these tasks. If there is a group of people who conduct this work in the company, discuss with them how much time is required (this will be a factor of how many externals there are, how many people need training, how complex the system is and how different the look and feel is compared to other systems the users are familiar with). In our example, we expect that 1/4 of the allocated time (or 30 days) should remain in the IS estimates to provide the support for Training and Manuals.

The User task time for Manuals and Training has been reviewed and an estimate of 100 days for each has been determined. Make sure that an assumption is created indicating the number of people to be trained, the timeframe, and the fact that the Users are conducting this and not IS.

We also need to estimate for the User involvement in the other phases of the project where they do not eliminate the IS time. We need to add substantial time to General Design where the Users are actively participating in defining the requirements and in Acceptance Testing where they are conducting the test with support from IS. We also need to add a moderate amount of time to Detailed Design, Programming and System Testing primarily to answer questions. Let's update our spreadsheet.

Synthesized version of the overall project total			
PHASE	Original Project Total	New Project Total	User Required Days
General Design	183	183	90
Detailed Design	366	366	20
Programming	832	832	20
System Test	283	283	25
Acceptance Test	183	183	120
Manuals	128	30	100
Training	128	30	100
Project Management	272	272	
Technical Architect	160	160	
Conversion	0	0	
Facilities Install	0	50	
Total	2,535	2,389	475

Spreadsheet A.3

Now you have an overall estimate of the effort. However, it is still an estimate, and more work is required to arrive at the estimate you can commit to.

A.3 Review Deliverables Required

At this point in the planning exercise, we want to review the overall scope of the project from a deliverable perspective. For simplicity sake, let's say our project requires the standard deliverables and the standard sections for these deliverables. In this case, we do not need to reduce the effort, just leave it as is.

A.4 Create High Level Gantt (phase level)

Ok, now comes the real fun part. How do you take the estimates, and decide how many people are required for how long. How is this done? As mentioned in Chapter 6, you can use a "square root method" to determine initial staffing and durations. How would this be done if this method did not exist? Well I'll explain how the thought process should work when doing this, so that without the square root method you would still be able to craft a proper plan.

In constructing our Gantt chart, use 240 man days per year, or 20 days per month. This allows for vacations, sickness and other non-project specific tasks. To specifically accommodate the statutory holidays, we will imbed 5 days for every six months of duration during the phase that spans the six month boundary.

Ok, let's look at each of the phases separately and determine how many resources we should use and what the duration should be.

General Design

Starting in the General Design phase, look at the number of overall man days and begin a chart that depicts the number of resources and the duration it would take them. Begin with one person and work upward. In our example above, on spreadsheet A.3 we have 183 man days for the General Design phase. Decomposing that into people time we have:

Number of People	Elapsed months to complete
1	9
2	4 1/2
3	3
4	2 1/4
5	1 3/4
6	1 1/2
7	1 1/4

In reviewing this it is easy to discount the one person option as being too lengthy in duration. Especially in General Design where there are Users to interview and a consistent coherent system to define, you need a reasonable duration. I would also rule out the 5, 6 and 7 people options because of this. This leaves us with two, three or four. Keep in mind we are creating the ideal plan now, so don't be overly aggressive. I would rule out the two because it is too lengthy for the amount of work. I would rule out the four as not as ideal as the three person option since it has more people and it is best to use as fewer people if possible.

Three people for three months - good duration, not too many people to be disruptive. On a sheet of graph paper, draw a box for the General Design phase, use 1 box on the graph paper for each week.

Detailed Design

Now let's look at the Detailed Design phase. Again, referring to spreadsheet A.3, we have 366 man days for Detailed Design. Decomposing that into people time we have:

Number of People	Elapsed months to complete
1	18
2	9
3	6
4	4 1/2
5	3 1/2
6	3
7	2 1/2
8	2 1/4

The first two options of one and two people clearly takes way too long, and are immediately eliminated. The last two options of 7 and 8 people would cause too much interaction, eliminate them. Interaction is the key, so consider how much interaction there is. Some of the design work will be data base and architecture related, but the vast majority will be program design. At the program design level, how much interaction is required between the program designers? Usually not too much. However, there will be lots of interaction between the architect, the data base designer and the program designers. So, the team should still be kept small. Three people is small, but six months is too long, eliminate it. We are left with 4, 5 or 6 people. Assume that one person is the technical architect and the data base designer, while the rest are program designers. Does it make sense to have 3, 4 or 5 **program** designers? Five program designers is too many, therefore eliminate the six person option. Either three or four program designers would work.

At this point I'd go with the option of four people (three program designers and the technical architect) for 4 1/2 months since it is fewer resources, but not an excessive long duration. On the graph paper that has the General Design box, create another for the Detailed Design phase with the duration of 4 1/2 months.

You should see that the Detailed Design phase now starts in the fourth month and ends in the eighth month. As mentioned earlier, we need to ensure we have accommodated the statutory holidays. We need not be totally accurate as to when they occur, since we may not know the starting month. We do know that there will be at least 10 of them every calendar year. At this point it is accurate enough just to slot time in the plan for them, and when the detailed plan is done it can slot these to their exact dates. So, add one more week onto the Detailed Design box to accommodate for 5 of the statutory holidays in the first six months of the project (cross-hatch this time to indicate it is for statutory holiday time). Your timeline should now show 7 3/4 months duration.

Programming

Referring to spreadsheet A.3, we have 832 man days for programming. However, remember in our case we have 32 days that we have planned for the Technical Architect to accomplish during the Programming phase, so we need to subtract that out from our total, leaving 800 man days. Decomposing that into people time, we have:

Number of People	Elapsed months to complete
1	40
2	20
3	13 1/2
4	10
5	8
6	6 3/4
7	5 3/4
8	5
9	4 1/2
10	4
11	3 3/4
12	3 1/2
13	3 1/4

Clearly, the first three options are way too long in elapsed time, so eliminate them. Thirteen people for 3 1/4 months is too many people for too short a time, as is 12 and 11 people. Eliminate these options. We are left with 4 to 10 people. Again, let's consider the technology. Is it known to the programmers? Is it an easy or difficult system? Are there a lot of difficult interfaces that need to work? All of these questions will direct you to the right timeframe. The more risk there is the longer duration you will want for this phase. Let's say our system is known to most, medium complexity and has only a few straight forward interfaces that have been programmed before for other systems. Basically, it is a system that has few potential major problem areas. Ten man months and eight man months is excessive for this system. Eliminate these options. A gain of 1/2 of a month by adding one person to a 9 person team doesn't seem like a good option, eliminate the 10 person option. Now we are left with the following:

Number of People	Elapsed months to complete
6	6 3/4
7	5 3/4
8	5
9	4 1/2

Again, going for the "best" plan now, I would opt for the six person team, but be willing to alter this to 7 or 8. I wouldn't want 9 people for 4 1/2 months as this is too many people for too short a duration. Let's stick with the 6 person team. Add a programming box to the graph sheet for 6 3/4 months.

Once again, the Programming phase has crossed over another 6 month duration so we need to add time in for statuary holidays. Add 5 more days to the duration of the Programming phase to accommodate these.

System Testing

Referring to the spreadsheet A.3, we have 283 days for System Testing. System Testing can be broken down into two major components - the preparation of the test cases with expected results and the execution of the tests (and correcting bugs). Usually, there requires fewer people to create the test cases than it takes to execute the test. A rough rule of thumb is that the test case development takes 20-30% of the effort and the execution takes 70-80%. The first component, the building of the test cases, should always begin before development is complete so that once development is complete, testing can begin immediately. However, the amount of overlap should be kept to a minimum - 3 or 4 weeks in our case is good in relation to a 7 month development schedule. The key here is that you don't want to re-work some test cases because you are too close to the time that they are being programmed to be certain of no changes to the specification (controlled by change requests).

Now let's review the two parts separately. The estimate for completing the test cases, using 20-30% of the total would fall into a range of 57-85 man days. To accomplish this in three weeks would take 4-6 people. To accomplish this in four weeks would take 3-4 people. Again, looking at the timeframe of seven months, I would opt for the 3 people for one month. Looking at the effort as well, 100 programs need test cases (as well as the other tests of performance, reliability, etc). Can 3 people for 20 days each reasonably accomplish this? That is 60 man days of effort. It seems well within the range of being "do-able". It is not overly important to be exact here, test case development can proceed with testing, or alternatively some test cases could be executed before development is complete. The latter becomes more complicated since "new" environments need to be created and controlled.

Now, we need to look at the execution part. We have 223 man days left to accomplish this (283-60). Decomposing that into people time, we have:

Number of People	Elapsed months to complete
1	11 1/4
2	5 1/2
3	3 3/4
4	2 3/4
5	2 1/4
6	2
7	1 1/2
8	1 1/2
9	1 1/4

Again, technical complexity can affect the duration. Having many interfaces and many technical complex sub-systems would entice the Project Manager to have a longer duration than normal in order to be able to react to problems. In our case we have a "normal" system as indicated before. So, in light of that the first two options are way too long and can be eliminated. The case of 3 people for almost four months is too long with too few people. The last two cases provide less than six weeks to test 100 programs and involves a lot of people. They can be eliminated. We are left with between 4 and 7 people for 2 3/4 to 1 1/2 months. Let's think about the team composition. We need people to enter and record the tests, and we need programmers to correct problems. We had a team of 6 programmers in development. We would want some of these to continue on into testing, while freeing others to move onto other projects. Two or three programmers would be appropriate, and we would need at least one analyst type but probably two. So, looking at it this way we have a minimum of 4 people and possible 5. With 5 people we can get it done two weeks quicker than with four people. With six people we gain 1 week. Seven people reduces the elapsed effort down to six weeks. This would be difficult to accomplish - everything would have to go **real** smooth.

Either the 5 or 6 person option would work. At this time being conservative I'd opt for the 5 person option and keep the extra week of elapsed time. On the overall graph, add a box that overlaps with Programming phase for four weeks, and then lasts nine more weeks and label it System Testing.

Acceptance Testing

Referring to spreadsheet A.3, we have 183 man days. This time to **support** the end user with their testing. It includes assisting in the creation of test cases and expected results, executing the tests and fixing the bugs. It is important to understand and have the User understand that this is **their** test; they own it, and the IS department is just assisting them. The types of IS resources required will be analytical to help with test cases and explaining the system, and programmers to fix the bugs.

As in the System Test, there are two components that should be separated out - the test case creation and the execution of the tests and correcting problems. The former should begin during System Testing and complete just prior to the end of System Testing. We don't want to overlap too much with System Testing otherwise we may get caught in a rippling affect - problems in System Testing changing acceptance test cases that have been created. Generally, you should not exceed 1/2 of the duration of the pure System Test timeframe. In our case, we shouldn't overlap more than one month.

Typically, the Acceptance Test involves fewer tests and fewer kinds of tests (performance may only be done at system testing and results provided to the user on their request). Also, fewer bugs should be found that require correcting and retesting. This means that the overall duration of the Acceptance Test should be shorter than the System Test. As a rule of thumb, the duration should be around 1/2 to 2/3 of the system test. If we look at the System Test, we have nine weeks of duration, 2/3 of that is 6 weeks, and 1/2 is 4 1/2 weeks. Rounding up to a full week, we have 5 or 6 weeks. Note, as the duration of the System Test increases, the ratio tends to gravitate to the 1/2 criteria. For instance on a six month System Testing, three months of Acceptance Testing is more reasonable than four. For our system, being conservative, I would opt for the six weeks.

Now, let's look at the number of resources. Assuming we start developing the test cases with three weeks of overlap (being conservative again), we should need less time than for System Testing which was 3 people for four weeks or 60 man days. Half of that time is appropriate given that we are supporting the Users, not doing all of the test cases. Therefore, for 30 days effort for three weeks, we would need 2 people to support the Users for this period. Reducing the overall estimate of 183 by 30 we have 153 man days. If our duration is six weeks, this would mean we would need just over 5 people to support this effort. This is a lot of support for a short period of time. In fact it is the same amount of support that we have during System Testing. We should require less support at this time since many of the bugs should have been corrected. However, at this point in time let's go ahead with it leaving it at six weeks. Add three weeks of time overlapped with System Testing and six weeks after that to our graph.

Note, we've now gone another 6 months, so we need to add another week to accommodate the holidays/vacations, etc. Add one more cross-hatched box

Other Phases

All of the other phases - Project Management, Training, Manuals, Conversion, Technical Architect and Facilities Install will fit within the overall duration of our current Gantt chart. The only case where this may not be true is in Conversion where the conversion may be larger than the project that you are developing. But before adding in these phases, let's look at the overall plan see what we have for a duration, and see if we can improve on it. The overall Gantt as it stands now is depicted below, on Gantt A.1.

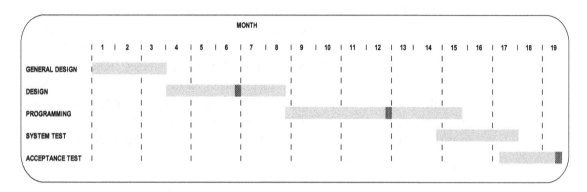

Gantt A.1

A.5 Tune the High Level Gantt

In reality and in practice, before adding in the "other" phases above, you should review the overall duration and see if you have been overly conservative, or if the overall timeframe is within management's' expectations. So, let's go back and review our Gantt chart. The total duration comes out to 18 3/4 months. Keep in mind this is our basic "risk free" project plan. We have used optimal resource decisions and been conservative throughout. Let's say there is some pressure to decrease this timeframe. How would we go about this? Well let's review each of the phases and see if we could add a resource or two, and what the affect would be on the duration.

General Design

First, the General Design phase - 3 people for three months. Adding a fourth resource cuts the timeframe to 2 1/4 months - a decrease of three weeks. This is lots of time, but what would be the ramifications? Well, how many distinct user areas are there to review, and how much of the analysis will be single threaded or have to be reviewed by two people? Let's assume there are two key users and most everything will have to go through them. Adding another resource to the analysis team may not have a positive effect, but wind up creating a bottle neck with the two users. Let's keep this estimate as is, for our example.

Detailed Design

Next, the Detailed Design phase. We have 4 designers for 4 1/2 months. Adding a fifth designer decreases the elapsed time by one month. Adding another person to this size design team and for a duration of 3 1/2 months is reasonable. Adding another resource to six designers gains only two weeks, and 6 designers is really too many. Therefore, we will add just one more resource.

Programming

We have 6 programmers for 6 3/4 months. Adding a programmer reduces the elapsed time by one month. Seven programmers for 5 3/4 months. This is entirely reasonable. Adding yet another one (to 8 programmers) reduces it a further three weeks. Eight programmers for 5 months. Again it is reasonable, but I would only go this way if there was real pressure to do so. This is because once we decrease this phase, overlap system testing's test creation by one month and begin User Manuals, we have an awful lot going on that becomes dependent on the application being programmed. The Users who are doing the manuals will want to know how transactions work and these transactions may not yet be programmed, and could result in some changes to the specifications from the General Design phase. Nevertheless, for the purposes of this example let's use the 8 programmers for five months. We will see some of the dependency problems later.

System Testing

We have five people for 9 weeks elapsed (not counting the overlapped part). Adding a sixth person decreases the timeframe to 8 weeks. This is not worth the extra risk.

Acceptance Testing

We currently have six weeks for 5 people. Decreasing the timeframe to five weeks adds one more person. However, this phase is driven by the Users, and by the number of test cases needed to complete the test. Can it be done it five weeks instead of six? The answer is yes, but it will put more onus on the Users. Will they be able to accomplish this? Let's say the answer is yes. In reality you would need to see what kind of a commitment the User is prepared to provide here, and the User would probably ask how much time you think is required of them, so be prepared. However, even if we do decrease the timeframe, we should not need the sixth person, and I would even argue we don't need 5 people, but let's remain conservative and keep 5 people for five weeks.

Now, if we look at the overall duration depicted on Gantt A.2, it is 15 1/2 months, down from 18 1/2. But we aren't finished yet. We need to add in the Training and Manual time in order to ensure we can properly handle the need for IS resources during this time period.

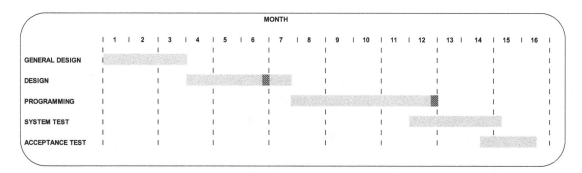

Gantt A.2

Manuals

For the Manuals we have 130 man days in total, with 30 of that used up for consultation. Of the remaining 100 days we have the following decomposition:

Number of People	Elapsed months to complete
1	5
2	2 1/2
3	1 1/2
4	1 1/4

Ideally one person should write the manual, but 5 months takes too long.

Two people is the next best option. Two and half months is reasonable. Let's go with that.

Training

For Training, there are two components that need consideration. The first is the preparation of the training materials and training scenarios, and the second is the actual training. It also makes sense that the individuals creating the User Manual create the Training Manual and do the training (although not required). The preparation part usually represents 70-80% of the time for training. Using this figure and the 100 days left in the training budget, the preparation time would fall into a range of 70-80 man days, and the execution the remaining 20-30 days. If we use the same two users who developed the manuals, the elapsed time for the preparation would be eight weeks (at 80 days) and the training itself would be two weeks (at 20 days). Again, sometimes these are affected by the number of users, their whereabouts and the levels of training required (how many different courses are needed). But let's say the above timeframes suit the Users.

Now add the Manuals and Training into the plan. The Users should all be trained before the Acceptance Testing is complete, but as near to the end as possible so that the training is still fresh. Some Users will need minimal training earlier in order to conduct the Acceptance Testing, but primarily they are executing from scripts, and the scripts can be generated from the User Manuals in conjunction with business scenarios familiar to the end User. So if we work backward from the end of the plan, leave one week at the end of Acceptance Testing. Moving back from there, we have two weeks of Training, and before that, eight weeks of User Training preparation, and before that ten weeks of User Manual preparation. This would put the start of the User Manual preparation about 1 1/2 months before Programming was complete and a couple of weeks before System Test case development started. This is reasonable. We now can add on Project Management for the duration, and a Technical Architect for the duration starting at Detailed Design.

A.6 Create Work Plan for First Phase

Now we would normally take the General Design phase and create a detailed spreadsheet of tasks and resources, and then map these onto a Gantt chart and ensure we have a smooth first phase, taking into account the critical path, the resources required, and trying to ensure everyone is contiguously busy. However, since we have a substantial amount of General Design time, I won't do this for our example. Refer to Section 7 of this document for the specific technique and expected output from this effort.

A.7   Create Low Level Gantt (phase level)

We are now ready for the final step in developing the high level plan - allocating resources to this Gantt and ensuring we haven't used the same resource in two places, or arbitrarily decided to get more programmers for System Testing and Acceptance Testing.

So take each phase and map out resource types who would do the job. For General Design, we need 3 people for three months. These would be analysts. On a separate graph sheet draw three lines for three months duration and label them A1, A2 and A3, for analyst 1, 2 and 3. Let's also add in a TA resource for about 5% of this effort, or about 10 days (183 days * 5% = 9.1 days).

In Detailed Design we need 5 designers including the overall technical architect, for 3 1/2 months. It would be reasonable to expect that the analysts have the skill set required to complete program design. Therefore, extend the three analyst lines for 3 1/2 months. Add a TA line and beginning at the start of Detailed Design, extend it for 3 1/2 months. Also, add another line for a designer, labeled D1, and extend it like the TA's.

In Programming, we need 8 programmers plus the TA, for five months. Extend the TA line for five months. Create eight more lines for programmers 1 through 8, label them PA1 to PA8 and extend them for five months.

For System Testing, we require 2 analysts and 1 programmer for developing the test cases and 2 analysts and 3 programmers to conduct the tests. The test case development was for one month, overlapped with programming, and the execution was for nine weeks. So, for two of the analysts (A1 and A2), draw a line one month before programming is finished. Now, we have a problem. We need a programmer to do test cases, but they are all busy programming. For now, create a PA9 position and draw a line for one month, starting one month before programming ends. Now, for those three resources and for PA1 and PA2, draw a line for nine weeks starting at the end of programming and label this System Testing. The TA is here throughout this phase so extend his line for the duration of system testing. OK, system testing is done.

For Acceptance Testing, we need 2 people for three weeks to help with the test cases and 5 people for five weeks to help conduct the test. The first part is overlapped with System Testing. We should have one analyst and one programmer. Since both other analysts are busy, we need to use A3. For A3 and for PA3, draw a line for three man weeks starting three weeks before system testing ends. Then, for A2, A3, PA3, PA4 and PA9, draw a line for five weeks. The TA is here throughout this phase as well, so extend his line to the end of Acceptance Testing.

Now, let's add in the Manuals and Training that the User will complete (to get the timeframes). Working backwards, training was for two weeks starting three weeks before Acceptance Testing ended. Draw a line for Users, depicting this. Moving backwards, extend this line eight weeks for the Training preparation. And finally, moving backwards again extend this line back ten weeks. Ok, now we indicated we would have 30 days to support Manuals and Training. It makes sense that part of this could be analyst time and part programmer time. For the analyst time, A3 has enough time to accommodate all 30 days. Use PA4 for all 30 days of Training assistance.

Finally, let's add in the Facilities Install time. In discussions with the technical services group, they have indicated we need two man weeks of effort for ordering and acquiring the equipment, two man weeks for installing the equipment for development, and two weeks to install the equipment for production. Now, we have to determine when these events need to occur and add these tasks to our Gantt chart. We will assume we need the development equipment during the design phase and the production equipment for System Testing. Therefore, add a line for FE1 during the General Design phase for one month, and another for his helper for the last two weeks. Let's also add two weeks for both just prior to System Testing commencing. In reality it is more complicated since sometimes the development machines will also be the production machines and the moving from the development to the production environment has to be done carefully.

The Gantt chart should now look like Gantt A.3 on the following page.

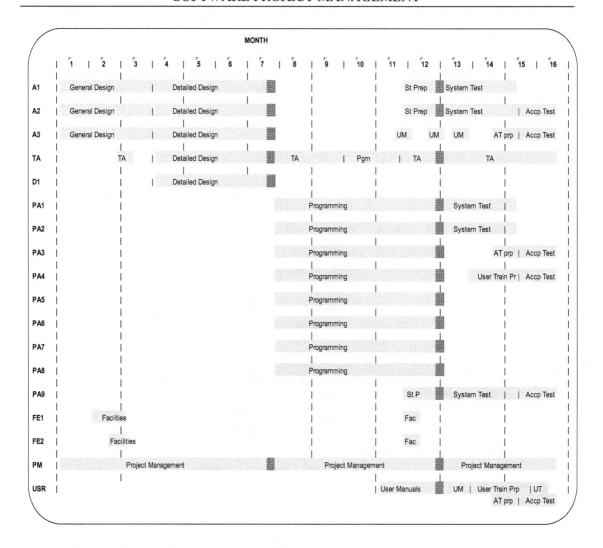

Gantt A.3

A.8 Tune Low Level Gantt chart

Now, we have a plan. Or do we? If you look at the Gantt chart, there seems to be some problems. The problems are:

1) We utilize D1 for only 14 weeks.

2) A1 has a 4 month gap between Detailed Design and System Testing.

3)   A2 has a 4 month gap between Detailed Design and System Testing.

4)   A3 has a 4 month gap after Detailed Design, as well as gaps after that and before Acceptance Testing.

5)   PA9 has not programmed anything, but comes on for System Testing, and Acceptance Testing.

6)   There is a gap between Programming and Acceptance Testing for PA3.

7)   There is a gap between Programming and user training support for PA4.

Well, let's try and solve these problems.   In order to show this last transformation we will create another Gantt chart.   Normally we would just adjust the one we are working on.   So create the new one (in pencil) and re-draw all components.

1)   The designer D1 could be a senior programmer who has the ability to do program design and program.   Let's use PA1 as a designer and as a programmer analyst.   Assign D1's detailed design time to PA1.

2)   Normally, if the IS staff are fully responsible for the project, they also write the user manuals.   In our case the end User is responsible.   But, during Programming, we will still require the knowledge of at least one of the analysts who worked during the General Design.   Therefore we will add Analyst A1 full time to the project by extending A1 from the end of Detailed Design to the start of System Testing.   During this time they will act as a team leader and assist in the documentation creation, including plans.   On the Gantt label this time as PM (for project management).

      Note that it is only acceptable to add project management time if the person is really going to participate as a team leader.   This can't be used as a contingency bucket to be allocated to anyone that you wish to keep on the project.

3) Let's re-adjust A3's time to others:

i) Reassign 15 days of A3's assisting with User Manuals to A1 (prior to System Testing, decreasing the amount of Project Management time we added in point 2).

ii) We have already designated PA4 as having the ability to assist Users. Let's give him the 15 remaining days for User Manuals.

iii) Lets add the 5 weeks of Acceptance Testing and 3 weeks of Acceptance Test preparation to A1. This reduces A1's system test time by 15 man days. But we still have the other analyst, A2, and A1 is still around to answer questions.

4) PA9 has 20 days for System Test preparation. Let's give PA9's System Test preparation time to PA1. How can we do that - we have just deleted 20 days of programming time! You are right. But I did think of that and decided there were two approaches:

i) Add 1 week of to the elapsed programming schedule. This has the effect of adding 40 days to the programming effort, and then takes 20 days away from PA1. The net effect is an additional 20 days (assuming I want all resources to end at the same time, and I do).

ii) Review the workload of the TA and ask whether he could absorb 20 days of programming. Well during programming, he has 100 days of work, and currently has 32 days of actual programs to write. I would feel comfortable in adding 20 days more, especially since they are not difficult and he will probably write them quicker than the 20 days allocated (since the estimates are for average programmers and the TA should be much better).

Given the two choices, I chose the latter. So let's allocate 20 more days of the TA to the programming effort and label it as programming time.

5) Let's complete the cleanup of PA9.

i)   Assign the 25 days of Acceptance Testing to PA1.

ii)   Assign PA9's System Test time to PA3. However, we
        have 15 fewer days left for PA3 than we need. Assign the
        remaining 15 days to PA5.

6)   One final adjustment and we are finished. There are too many
        Acceptance Test programmers (the same number as in System
        Testing. Let's get rid of PA1's Acceptance Test time.

The end result of these changes is depicted on Gantt A.4.

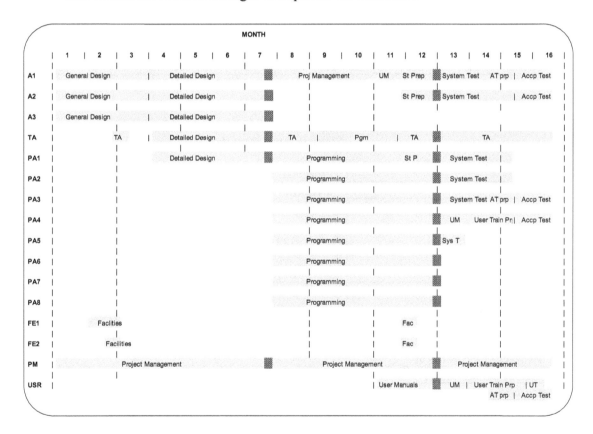

Gantt A.4

We now have a plan we can bring forward to management. So, finally, let's add
up the resources by resource type and by phase. Spreadsheet A.4 depicts the
total by resource type and phase, while spreadsheet A.5 depicts the resources by
phase and shows the differences between the original totals, the new totals and

the used totals.

Man days on low level on low level Gantt chart															
	PM	A1	A2	A3	TA	PA1	PA2	PA3	PA4	PA5	PA6	PA7	PA8	FE1	FE2	TOTAL
Gen Design		60	60	60												180
Dtld Design		70	70	70	70	70										350
Programming					52	80	100	100	100	100	100	100	100			832
System Test		50	65			65	45	30		15						270
Accept Test		40	25					40	25							130
Manuals		15							15							30
Training									30							30
Project Mgmt	300	65														365
Tech Arch					128											128
Conversion																0
Facilites														30	20	50
TOTAL	300	300	220	130	250	215	145	170	170	115	100	100	100	30	20	2,365

Spreadsheet A.4

	Consolidated Man days from Gantt			
PHASE	Original Project Total	New Project Total	Used in Plan	User Required Days
General Design	183	183	180	90
Detailed Design	366	366	350	20
Programming	832	832	832	20
System Test	283	283	270	25
Acceptance Test	183	183	130	120
Manuals	128	30	30	100
Training	128	30	30	100
Project Management	272	272	365	
Technical Architect	160	160	128	
Conversion	0	0	0	
Facilities Install	0	50	50	
Total	2,535	2,389	2,365	475

Spreadsheet A.5

Note the differences between the three totals. Had we just used the percentages without thinking about the project in more detail, we would have predicted roughly 150 more days for the IS staff then we actually believe to be the right number. Although the difference between the total in the "New" and the "Used" columns is less significant (in this case) it shows the progression from the percentages to the final figures. This chart is especially useful and significant to IS Management as it shows the progression from the standard percentages to the "tuned" numbers. It is important to have footnotes for the phases that have changed significantly from the percentages, explaining why this was the case, so IS management can understand why the numbers have been adjusted.

APPENDIX B:
PERCENTAGE ALLOCATION

B.1 OVERVIEW

Once the total number of man days for programming is determined, there is a standard percentage method that is **intended** to provide a baseline **estimate** for all of the phases and major tasks. These percentages are based on the definition that programming is 100% and the others are a percentage of this. Let's look at the percentage allocation:

PHASE	Percent
General Design	22%
Detailed Design	44%
Programming	100%
System Test	34%
Acceptance Test	22%
Manuals	16%
Training	16%
Project Management	34%
Technical Archtect	20%
Conversion	0%
Facilities Install	0%

Spreadsheet B.1

Thus, General Design is 22% of the programming effort, and System Testing is 34% of the programming effort.

What does this all mean, and why is there **any** correlation between any of these activities? Why are there no percentages for Conversion and Facilities Install?

Firstly, from a historical perspective, these percentages were obtained from actual projects statistics, over the course of 10 years. However, this really does not answer the question as to why there is a correlation.

Let's backtrack for a moment. What information do we know with any amount of certainty? Well, if we have gone through the externals and estimated the programming effort, we should know:

- The number of externals,

- The complexity of each external, and

- The number of man days required to program each external.

As it turns out, the first 7 phases on the percentage chart above are really driven by the number of externals, although the first five have a higher correlation. Thus, if there were 50 externals to program, there is an associated amount of work to document the externals in General Design, to design them in Detailed Design, to test them in System and Acceptance Test, to train the users in Training and to document them in the Manuals phase.

So why are there no percentages for Conversion and Facilities Install? Firstly, all of the other phases and tasks are **always** required in **every** project. However, not every project requires Conversion or the installation of facilities. On the other hand, sometimes conversion of data is a larger project than the system itself. There does not seem to be any rationale in assigning any arbitrary number to these phases, but rather ensure they are thought of, even here at a high level.

As mentioned above, the phases that have a strong correlation to the percentages are the General Design, Detailed Design, Programming, System Test and Acceptance Test. The others are less reliable.

Let's look at all of the phases/tasks separately and understand the effort, and the correlation to the programming effort.

B.2   GENERAL DESIGN

For General Design, we have 22% of the effort required to program the system. How accurate is it? How can I verify, rationalize, and finalize the figure? What might cause variances?

Well, as for all of the percentages, this percentage gets us into a "reasonable" range. How can that be? Well, let's look at an example that has 500 man days of programming, and there are 50 externals. From our percentage of 22%, we derive that General Design is 110 man days (22% * 500). Now how do we validate this number? Well, let's look at what is involved in analysis, and conduct a task estimate. From the methodology guidelines manual, the major activities are the following:

1)   Develop Horizontal Prototype

2)   Develop the Functional Specifications

3)   Revise Detailed Logical Data Model

4)   Update Divisional Data Model

5)   Develop Implementation Strategies

6)   Document Systems Acceptance Criteria

7)   Document Technical Requirements

8)   Develop Vertical Prototype

9)   Update the Business Case

10) Prepare a Detailed Plan for Detailed Design and a General Plan for the Remaining Phases

11) Review and Finalize General Design Deliverables

Now let's do a task estimate for the General Design phase to ensure we know what we have to do and how long it will take. So, using the above activities, create a spreadsheet of these activities along the side and resources along the top. For resources, list the Project Manager, two System Analysts, a Data Administrator and the Technical Architect. Spreadsheet B.2 depicts the resources and activities.

Activity	PM	SA1	SA2	DA	TA	TOTAL
Develop Horizontal Prototype						0
Develop Functional Specs						0
Create/Revise LDM						0
Update Divisional Data Model						0
Develop Impl. Strategies						0
Document System Acceptance						0
Document Technical Reqts						0
Develop Vertical Prototype						0
Update Business Case						0
Develop PCD for Dtld Design						0
Review and Finalize Delivs.						0
TOTAL	**0**	**0**	**0**	**0**	**0**	**0**

Spreadsheet B.2

Now, let's look at each of the tasks separately, and determine an estimate and the degree of correlation this task has to the overall programming effort. Assume that there are 50 externals and there are no unusual technical or implementation requirements.

Develop Horizontal Prototype

This task involves the review of all screens and reports with the User, and the development of each of these using an automated tool. The task of reviewing the requirement with the User and prototyping an external usually takes a couple of hours for each external.

The effort to complete this is clearly related to the number of externals that need to be prototyped. This task has high correlation to the programming effort.

There are 50 externals. A reasonable estimate to develop a screen or report sample is 2 hours. That gives us 100 hours, or about 13 man days ($100/7.5 = 13.333$). Divide this time between the two system analysts.

Develop the Functional Specifications

This task involves creating of the detailed specifications for WHAT the system is expected to do. Specific and detailed documentation is created for each component of the system. This task involves preparing for an interview (or JAD session), conducting the interview, documenting the results and reviewing the results with the User for validity.

Clearly there is a high correlation to the programming effort. For every external that was estimated to get the programming man days, there needs to be time spent to define the external. The amount of time also correlates. If there were only one external and the estimate to program it was 10 days, then there would be 2.2 days of General Design. If the one program was estimated to be 30 days, then there would be 6.6 days. Clearly, the more complex and harder to program, the more time it will take to specify the requirement.

This task generally takes between 10 and 14 hours per external, depending on the types of externals. Let's be aggressive and use the 10 hours for our estimate. We have 50 externals times 10 hours each, gives us 500 hours, or about 66 days. Divide this between the two system analysts.

Create/Revise Detailed Logical Data Model (LDM)

This task, is somewhat bound by the number of data elements (attributes) and the processing of them. However, creating a logical data model can continue to go on for longer than can be expected (forever?). From the previous phase (Project Definition), we should either have a High Level Data Model, or a list or grouping of the data that the system requires or enacts upon. During the process of doing the Functional Analysis, new data or changes will be found by the systems analyst. The time allocated for this task should only be for the Data Administrator to assist the Systems Analyst in determining where and how data should be placed, and the final packaging of the deliverables. We should be able to fix the amount of time that is spent on this task.

The amount of effort here depends on the amount of data in the system, which correlates to the number of screens and reports that are to be developed. Since screens and reports tend to have varying number of data elements on them, there is not a high correlation.

This task should not be long in duration since it is updating the LDM created in the Project Definition phase. Expressing a meaningful estimate without any data is difficult. Given the number of externals and the amount of data that is typically on any external, we could project that this task is **probably** 5 to 10 days. Anymore and we are probably doing too much work. For the purposes of our example let's use somewhere in the middle, 8 days. Allocate six of these days to the Systems Analyst and the remaining two to the Data Administrator.

Update Divisional Data Model

This task has no correlation to the programming effort. It is affected by how close the existing Divisional Data Model is to the data model for this project.

An estimate for this task is difficult without knowing the status of the divisional data model and the likely changes due to this project. In general, if the Divisional Data Model is fairly complete, one would expect only minor adjustments would be needed. An estimate of 1 to 5 days would seem reasonable. For our example let's use somewhere in the middle, 3 man days. Allocate these days to the Data Administrator.

Develop Implementation Strategies

This task has no correlation to the programming effort. The strategies are dependent on many factors that have no relationship to the effort to program a system.

The strategies to be explored and documented are "big bang", phased and Pilot implementations. The deliverable is **only** strategies and not detailed documents. This should be a **bounded** task. Five days per strategy should be adequate, with one day for assisting the business resources on their strategy. Divide the 16 days between the two Systems Analysts.

Document Systems Acceptance Criteria

This task has very little correlation to the programming effort. It is a simple task that involves specifying the criteria that the User will use to accept the system. Included in this task is the definition of how many scenarios will be tested, and their complexity.

This task should be done by a Systems Analyst and the User community and should take only one day for systems with less than 100 externals and an additional day for every 100 more externals.

Document the Technical Requirements

This task has very little correlation to the programming effort.

This task should be assigned to the Technical Architect, and should be bounded. There are a few variables such as the similarity of the hardware and software to existing systems, but for the most part, the task involves documenting the **requirements** of performance, backup, etc. This task should be in the range of 5 to 20 days. For our example let's use 10 days. This time will be added to the Technical Architect time, and not be part of the General Design estimate.

Develop Vertical Prototype

This task has very little correlation to the programming effort.

This task can be as very large or non-existent. The time should be allocated to the Technical Architect. This time will show up as under the Technical Architect time, not the General Design time. However, for completeness let's add 20 days to our estimate

Update Business Case

This task has a little correlation to the programming effort, in that the amount of effort to build the system is dependent on the size of the system. The larger the size of the system, the greater amount of work it is to create and update the business case.

The Users need assistance in completing the costs for the business case. The Project Manager should provide the required assistance from the IS side. This could be from 1 to 10 days, so let's allocate 5 days to the Project Manager.

Prepare a Detailed Plan for Internal Design and a General Plan for the Remaining Phases

This task has some correlation to the programming effort since the amount of effort to plan is dependent on how many tasks need to be planned and this is dependent on the number of programs that need specifications.

This is the Project Manager's task. The plan normally takes between 5 and 20 days depending on the size of the system. For this example let's use 15 days.

Review and Finalize General Design Deliverables

This task involves a final review of the General Design deliverables, usually concentrating on the Functional Specifications. Because the Functional Specifications have a high correlation to the programming effort, this task has a high correlation as well.

This review should include the Systems Analysts, the Technical Architect, and the Project Manager. Depending on the size of the project it could take between 1 and 5 days. Let's use 3 days in our example.

Spreadsheet B.3 depicts the effort for each of the tasks that we estimated.

TASK	PM	SA1	SA2	DA	TA	TOTAL
Develop Horizontal Prototype		6	7			13
Develop Functional Specs		33	33			66
Create/Revise LDM			2	6		8
Update Divisional Data Model				3		3
Develop Impl. Strategies		9	7			16
Document System Acceptance		1				1
Document Technical Reqts					10	10
Develop Vertical Prototype					20	20
Update Business Case	5					5
Develop PCD for Dtld Design	15					15
Review and Finalize Delivs.	3	3	3		3	12
TOTAL	23	52	52	9	33	169

Spreadsheet B.3

Ok, so how does the estimate of 169 days compare with our anticipated estimate of 110? Well, first of all we have to eliminate the time for tasks that are not to be allocated to General Design time. This includes the Project Manager's time and the Technical Architect's time. We are left with an estimate of 113 days for General Design.

We have proved nothing yet, other than our percentage estimate is reasonably close to our task estimate (one could argue that in an example such as this the numbers could be contrived to arrive at this). So, let's look at those tasks that have a correlation to the programming effort.

One could postulate that if there were a higher number of days spent in General Design on tasks that have a high correlation, then there would be a stronger basis for the percentage allocation accuracy.

The tasks with a high correlation to the programming effort are:

- Develop the Horizontal Prototype,

- Develop the Functional Specifications, and

- Review and Finalize General Design Deliverables

The total number of days for these tasks is 85 (don't include the TA or PM time for review). Dividing this by the total for General Design tasks only (113), we get the percentage of time that is spent on tasks with a high correlation to the programming effort. This percentage is 75.22%. Therefore, 75% of the effort is directly related to the number of externals. As the number of externals increased or decreases, the amount of time for 75% of the tasks will vary. Keep in mind some of the other tasks had a small correlation; this percentage is only for those tasks that were high in correlation.

B.3   DETAILED DESIGN

OK, so what about the correlation of the Detailed Design to the programming effort. Well, I won't go into as much detail as for General Design, but let's look at the major tasks in Detailed Design. They are:

- Document Overall Architecture Components
- Document Physical Environment
- Design Technical Components
- Create Program Specifications
- Design Physical Data Base
- Create Test Plan

- Create Test Environment
- Create Conversion Program Specifications
- Prepare Detailed Plan for Programming

Two of these tasks are handled outside of the estimate for Detailed Design - Create Test Environment and Prepare Detailed Plan for Programming. As for the remaining tasks, most of the time in this phase is spent on creating the program specifications and the conversion program specifications. Generally these two tasks will take at least 80% of the effort. These two tasks have a high correlation to the programming effort, since the specifications are for the programs.

It would be relatively safe to say that there is at least an 80% correlation to the programming effort. The more programs there are to write, the more time needed to develop the specifications for them.

B.4 PROGRAMMING

Clearly in the Programming phase there is a 100% correlation.

B.5 SYSTEM TEST

System testing has two primary goals. One is to test the functionality of the system is as specified in the Functional Specifications. The second is to test that the technical components as specified in the Technical Requirements document work as specified. Most of the tests in a system test are concerned with the first goal. It would be fair to say that in typical systems 90 to 95% of the effort in System Testing is related to the functionality testing and the remainder is for technical testing.

Clearly the functionality testing has a strong correlation to the number of externals and hence the programming effort. The more externals there are the more testing that is required.

However, one cannot say that every system that has 50 externals requires the same amount of testing. The type of system that is to be tested and the risk that is to be realized plays a role in the estimate. For instance, for a Police or Fire dispatch system or a Warrant system where loss of life is at stake, more testing may be required. Also, if the system is the lifeblood of the company, say a financial system, where small errors can result in millions of dollars of loss, additional testing may be required.

Nevertheless, it should be clear that there is a high correlation between the effort to test the system and the amount of time to program the system.

B.6   ACCEPTANCE TEST

The same argument that was stated for System Testing holds true here. Usually, there is even a higher amount of effort concentrated on the functionality of the system and less on the technical.

The same conclusion can be drawn for Acceptance Testing as was for System Testing - that there is a high correlation to the programming effort.

B.7   MANUALS

The development of manuals for the end User and for operations has some correlation to the programming effort. The operations documentation has little correlation, but the end user documentation has a higher correlation. This is because typically the user documentation consists of an overall guide to the system as a whole, and then, for each external that the user works with, there is a detailed explanation as to how to use it. Clearly the number of these externals and their complexity will affect the effort to complete this task.

The more time that is assigned to writing technical programs that the user does not see, the less the correlation becomes. For instance in an environment where a transaction is sent from the Agent's laptop to the mainframe for underwriting, there is considerable amount of programming to conduct the handshaking and ensuring the integrity of the data. All of this code is in the programming effort, but none of it requires documentation in the user manuals. It is for this reason that the correlation is not as great as the other phases (in the other phases it doesn't matter that the programs are technical underpinnings or user externals).

B.8   TRAINING

The same statements that were made regarding Manuals holds true with

Training. The more technical components, the less correlation. But, for every user external, there needs to be scenarios that use them and time spent training the end User to use them correctly. The more the number of externals and their complexity (which makes up the programming effort) the more training time is required.

APPENDIX C:
PROJECT PHASES

The intent of this chapter is to provide a brief overview of the phases that are used throughout this document in order to place these phases in perspective with the phases that the reader is familiar with.

There are 13 phases that are referenced in this book. While some of these are not "real" phases but a grouping of high level tasks, they are referenced as phases for easy referencing when referring to the components that need to be estimated and planned for. The 13 phases are:

1. Project Initiation

2. Project Definition

3. General Design

4. Detailed Design

5. Programming

6. System Test

7. Acceptance Test

8. Manuals

9. Training

10. Project Management

11. Technical Architect

12. Conversion

13. Facilities Install

C.1 PROJECT INITIATION

Project Initiation is the initial phase of any project. The objective of this phase is to clearly state the business problem or opportunity. Deliverables produced include a Project Objectives Document and a plan for the next phase (Project Definition).

C.2 PROJECT DEFINITION

The intent of this phase is to define the scope of the system by defining business requirements. Major tasks in this phase include conducting a Current Systems Review, defining requirements for the new system and developing a conceptual design of the new system. The conceptual design is specific to actual externals that will need to be developed, but their definition is at a high level. Some methodologies refer to this phase as the Requirement Definition Phase.

C.3 GENERAL DESIGN

The objective of this phase is to document the business requirements of the new system in terms that the User understands, and is detailed enough in order to accurately develop a plan. Typical tasks include screen and report design, definition of data, definition of processing logic and definition of technical requirements (such as security, availability and performance). The deliverables from this phase represent the agreed to specifications that now form the basis for all future change control.

Some methodologies refer to the tasks in both Project Definition and General Design simply as the Analysis Phase.

C.4 DETAILED DESIGN

Detailed Design's objective is to develop an optimum technical solution given the business and technical requirements defined in General Design. Typical tasks include sub-system definitions, program specifications, and data base design.

C.5   PROGRAMMING

The programs for all screens, reports and all other processes are coded and unit tested in this phase.

C.6   SYSTEM TEST

This is the developer's detailed test of the complete system. In encompasses many specific tests such as performance, functionality, security, and interfaces. It encompasses the test case preparation, test file preparation, testing and correcting of errors.

C.7   ACCEPTANCE TEST

This is the User's test of the system to ensure it meets the requirements as stated in the General Design (and any subsequent Change Request). It typically is concerned with the functionality, but may also test some technical aspects of the system.

C.8   MANUALS

This phase involves the preparation of all manuals that are to be delivered with the system. It may include such documents as an Operations Document, User Manuals, and Programmers' Handbooks.

C.9   TRAINING

This phase includes all preparation of training material and the actual training itself. It includes end-user training, operations training and maintenance training.

C.10   PROJECT MANAGEMENT

This phase encompasses all aspects of the tasks associated with managing a project.

C.11   TECHNICAL ARCHITECT

This phase is concerned with all tasks that are the responsibility of the Technical Architect.   This involves such items as architecture responsibility, technical reviews, development and implementation of a System Development Environment (SDE) and programmer oversight.   On many projects this phase may be assigned to only the Technical Architect, while on others it may be a team.

C.12   CONVERSION

This phase encompasses all aspects of Conversion.   This includes all analysis, design, programming, testing, etc that is required to convert the existing system. On some projects it may consist of a part time resource while on others it may consist of a team larger than the main project.

C.13   FACILITIES INSTALL

This phase deals with all aspects of installing facilities.   It includes the installation of the development facilities and the production facilities.

APPENDIX D:
ABBREVIATIONS USED

AA – Application Architect

AR – Accounts Receivable

AP – Accounts Payable

Ax – Analyst 1,2 , etc

BPR – Business Process Re-Engineering

BR – Business Representative 1,2 etc

BS – Business Sponsor

CR – Change Request

DA – Data Analyst

DBMS – Data Base Management System

DFD – Data Flow Diagram

DR – Decision Request

Dx – Designer 1,2, etc

EWA – Enterprise Wide Architecture

FEx – Facilities Engineer 1,2 etc

GL – General Ledger

IS – Information Systems

JAD – Joint Application Development

LAN – Local Area Network

OO – Object Oriented

PAx – Programmer Analyst 1,2, etc

PD – Project Director

PM – Project Manager

POD – Project Objective Document

RAD – Rapid Application Development

RFP – Request for Proposal

SAx – Systems Analyst 1,2 etc

SPA – Senior Programmer Analyst

SDE – System Development Environment

TA – Technical Architect

TOC – Table of Contents

TRO – Temporary Restraining Order

TSD – Technology Selection and Deployment

TSx – Technical Specialist 1,2, etc

UA – User Analyst

UC – User Co-coordinator

WBS – Work Breakdown Structure

INDEX

www.ingramcontent.com/pod-product-compliance
Lightning Source LLC
Chambersburg PA
CBHW080359060326
40689CB00019B/4073